Forest Ecosystems: Forest Structure and Biodiversity

Forest Ecosystems: Forest Structure and Biodiversity

Edited by **Lee Zieger**

New York

Published by Callisto Reference,
106 Park Avenue, Suite 200,
New York, NY 10016, USA
www.callistoreference.com

Forest Ecosystems: Forest Structure and Biodiversity
Edited by Lee Zieger

International Standard Book Number: 978-1-63239-343-2 (Hardback)

Printed in the United States of America.

Contents

Preface   VII

Forest Structure and Biodiversity   1

Chapter 1   **Deadwood in Forest Ecosystems**   3
Katarína Merganičová, Ján Merganič, Miroslav Svoboda,
Radek Bače and Vladimír Šebeň

Chapter 2   **Plant Diversity of Forests**   31
Ján Merganič, Katarína Merganičová, Róbert Marušák
and Vendula Audolenská

Chapter 3   **Advances in Molecular Diversity of
Arbuscular Mycorrhizal Fungi
(Phylum Glomeromycota) in Forest Ecosystems**   57
Camila Maistro Patreze, Milene Moreira and Siu Mui Tsai

Chapter 4   **Arthropods and Nematodes:
Functional Biodiversity in Forest Ecosystems**   85
Pio Federico Roversi and Roberto Nannelli

Chapter 5   **Vegetation Evolution in the Mountains
of Cameroon During the Last 20,000 Years:
Pollen Analysis of Lake Bambili Sediments**   109
Chimène Assi-Kaudjhis

Chapter 6   **Using Remotely Sensed Imagery for
Forest Resource Assessment and Inventory**   137
Rodolfo Martinez Morales

Chapter 7   **Composition and Stand Structure of Tropical
Moist Deciduous Forest of Similipal
Biosphere Reserve, Orissa, India**   151
R.K. Mishra, V.P. Upadhyay, P.K. Nayak, S. Pattanaik
and R.C. Mohanty

Permissions

List of Contributors

Preface

Over the recent decade, advancements and applications have progressed exponentially. This has led to the increased interest in this field and projects are being conducted to enhance knowledge. The main objective of this book is to present some of the critical challenges and provide insights into possible solutions. This book will answer the varied questions that arise in the field and also provide an increased scope for furthering studies.

Forests are much more than a bunch of trees, contrary to popular belief. They are complicated operating systems of interrelated biological, physical and chemical elements, whose biological features have constantly evolved to preserve itself. These complex combinations of climate, soils, trees and plants result in many varied types of forests throughout the globe. Although trees are the main highlight of forest ecosystems, the vast variety of other creatures and abiotic elements in the forests indicate that there is also a need to study about these elements like soil nutrients and wildlife and consider these compounds when forming management plans. This book does not only provide information about trees, like other books, but also gives an insight into other components related to forests, which are usually left untouched in other texts, even though they are as important as trees to maintain balance in ecology. The book discusses structure and biodiversity and examines researches done on ecological structures.

I hope that this book, with its visionary approach, will be a valuable addition and will promote interest among readers. Each of the authors has provided their extraordinary competence in their specific fields by providing different perspectives as they come from diverse nations and regions. I thank them for their contributions.

Editor

Forest Structure and Biodiversity

Deadwood in Forest Ecosystems

Katarína Merganičová[1,2], Ján Merganič[1,2],
Miroslav Svoboda[1], Radek Bače[1] and Vladimír Šebeň[3]
[1]Czech University of Life Sciences in Prague,
Faculty of Forestry, Wildlife and Wood Sciences, Praha
[2]Forest Research, Inventory and
Monitoring (FORIM), Železná Breznica
[3]National Forest Centre, Forest Research Institute, Zvolen
[1]Czech Republic
[2,3]Slovakia

1. Introduction

Until recently, deadwood was perceived as a negative element of forest ecosystems, that indicates "mismanagement, negligence, and wastefulness" of the applied forest management (Stachura et al., 2007). It was regarded as a potential source of biotic pests, mainly insects (Bütler, 2003; Marage & Lemperiere, 2005), to remaining trees in a forest as well as to adjacent stands (Pasierbek et al., 2007). The presence of deadwood was also seen as a threat of the spread of abiotic disturbances, e.g fire (Thomas, 2002; Travaglini et al., 2007). In managed stands, deadwood represented an obstacle to silvicultural activities (Travaglini & Chirici, 2006; Travaglini et al., 2007), and reforestation (Thomas, 2002). Considering forest workers and visitors, standing dead trees have been seen as a threat to public safety (Peterken, 1996; Thomas, 2002) that had to be removed immediately after they had occurred (Pasierbek et al., 2007). For these reasons, sanitary cuttings have been common forestry activities not only in managed forests, but also in protected areas (Pasierbek et al., 2007; Stachura et al., 2007). In Europe, the maintenance of "hygienic standards" of a forest through systematic removal of sick, dying, and dead trees has been a common practice for more than 200 years (Stachura et al., 2007). In traditional systems, nearly every piece of wood would have been utilised (Mössmer, 1999; Butler et al., 2002). While large deadwood was usually extracted from the forests during stand tending (Radu, 2007), small wood pieces and leftovers were often burnt (Travaglini & Chirici, 2006). This intense forest exploitation has led to a substantial decrease of deadwood quantities (Travaglini & Chirici, 2006).

Over the last decades, the perception of deadwood in forest ecosystems has gradually changed as the scientific research gained information about the functions of deadwood in forests. North American researchers were the first to recognise the importance of deadwood presence in forest ecosystems (Radu, 2007). Already in the first half of the twentieth century several authors (e.g. Graham, 1925; Kimmey & Furniss, 1943; Savely, 1939, as cited in Thomas, 2002) identified deadwood as an important habitat for wildlife. In 1966 Elton (1966) described the role of deadwood in forests as a critical habitat component for a great number of species. In America, researchers as well as forest managers recognised the importance of deadwood in the ecology of a forest as early as in the 1970s (Thomas, 2002). This triggered additional research to further expand knowledge about the role of deadwood in forest ecosystems. Since then, a number of publications have documented its importance particularly for biodiversity (Ferris & Humphrey, 1999; Humphrey et al., 2004; Müller & Schnell, 2003; Schuck et al., 2004), nutrient cycling (Harmon et al., 1986; Krankina et al., 1999; Lexer et al., 2000; Pasinelli & Suter, 2000), natural regeneration (Harmon & Franklin, 1989; Mai, 1999; Ulbrichová et al., 2006; Vorčák et al., 2005, 2006; Zielonka, 2006) and other processes.

Nowadays, deadwood is of interests not only to ecologists, but also to mycologists, foresters, and fuel specialists (Rondeux & Sanchez, 2009). It is increasingly recognised as an important component in the functioning of forest ecosystems (Vandekerkhove et al., 2009) and is becoming an integrated part of forest management (Marage & Lemperiere, 2005). This is proved by the fact that deadwood has been selected as a Pan-European indicator of sustainable forest management (Ministerial Conference of Protection of Forest Ecosystems [MCPFE], 2002). Deadwood is also one of 15 main indicators of biodiversity proposed by European Environmental Agency (Humphrey et al., 2004). Within the Forest Inventory and Analysis program in the USA, deadwood is an indicator of forest structural diversity, carbon sources and fuel loadings (Woodall & Williams, 2005).

Hence, the objectives of the presented paper are (i) to review the approaches for deadwood description, characterisation, and evaluation, (ii) to review the importance of deadwood for sustainable forest management.

Fig. 1. Deadwood as a part of natural forest ecosystem. (photo by J. Vorčák)

2. Definition and types of deadwood

Schuck et al. (2004) and Rondeux & Sanchez (2009) presented several definitions of deadwood that were gathered within the activities of COST Action E272, and thus showed the variability in understanding of this term, which mainly depends on the aim of the particular study (Rondeux & Sanchez, 2009). However, from a general perspective the term deadwood encompasses all woody material in forests that is no longer living including stems, or their parts, branches, twigs, and roots, but excluding deadwood parts of living trees. Hence, it includes both aboveground and belowground woody material (Harmon & Sexton, 1996). It can originate either from the natural mortality caused by senescence, competition, or disturbances (windthrow), or from silvicultural treatments (Rondeux & Sanchez, 2009).

The aboveground woody debris can occur as standing dead trees or shrubs or their partial remains, or as material lying on the forest floor (Pyle & Brown, 1999). Belowground material include dead woody roots and buried wood, which is very decayed woody detritus found in the mineral soil or forest floor (Harmon & Sexton, 1996). However, since belowground material is difficult to quantify (Schuck et al., 2004), it is only rarely accounted for in the studies; although its proportion may be particularly in managed forest stands significant (Debeljak, 2006).

As vary the definitions of deadwood between the studies, so do the types specified in individual works. While some authors (e.g. Atici et al., 2008; Fridman & Walheim, 2000; Christensen et al., 2005; Vacik et al., 2009) distinguish only two types of deadwood, namely standing and lying deadwood, other works use a more detailed classification with four or five deadwood components (Kirby et al., 1998; Schuck et al., 2004; Travaglini et al., 2007). The basic classifications distinguish between standing and downed or fallen deadwood, while the common separation limit is at a 45-degree angle (Harmon & Sexton, 1996; Rondeux & Sanchez, 2009). There is a clear difference in the decomposition process between the two types of deadwood and in the host species. While birds and lichens are almost entirely associated to standing dead trees, fungi and moss species primarily colonise lying deadwood (Stokland et al., 2004).

Standing deadwood consists of standing dead trees, snags, and stumps. Snags are defined as vertical pieces of dead trees. There exists a discrepancy in the understanding of snags and stumps between some authors. For example, Harmon & Sexton (1996) consider a snag any vertical piece irrespective of its height resulting from natural processes only, while Travaglini et al. (2007) require a snag to have a height equal to or greater than 1.3 m. Consequently, in the first work, a stump is defined as a short vertical piece resulting from cutting (Harmon & Sexton, 1996), but in the second paper, it is a piece shorter than 1.3 m irrespective of its origin. Lying deadwood includes downed dead trees, and lying deadwood pieces, which are often called logs. The specification of log parameters depends on a particular study. For example, Pyle & Brown (1999) define a log as a piece at least 1.5 m in length, while Debeljak (2006) considers a log any lying woody piece with length 1 m or more.

The most important size distinction is between coarse and fine woody debris (Harmon & Sexton, 1996) representing large and small pieces, respectively. The two categories are separated depending on the diameter at a specific point on a tree or a log. However, the threshold value varies from 0 to 35 cm (Cienciala et al., 2008). According to IPCC (2003), the border diameter is 10 cm. Harmon & Sexton (1996) found that below this diameter the decay rate increases exponentially, while above this diameter the decay rate decreases only slowly. Due to this fact, the decomposition process of coarse woody debris (CWD) can sometimes take up to 1,000 years (Feller, 2003) depending on wood characteristics (tree species, dimensions), climate characteristics (temperature and moisture, Woodall & Liknes, 2008) and the position on the ground (i.e. contact with the ground, Radtke et al., 2004). Since CWD persists a substantial time in the ecosystem, it is regarded as a more significant component of deadwood than fine woody detritus. Hence, most inventories do not account for fine woody debris and deal only with the components of coarse woody debris.

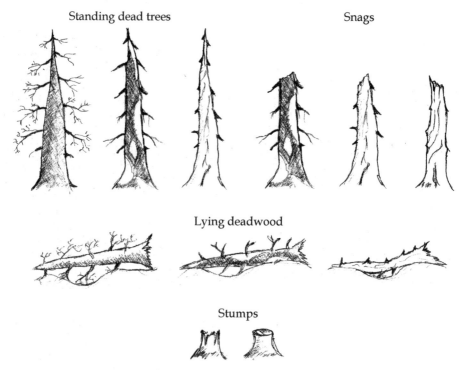

Fig. 2. Types of aboveground deadwood.

3. Deadwood assessment

Nowadays, deadwood is assessed in many countries of the world within national inventories, and through various forest research activities. However, there is no accepted standard for the assessment of deadwood (Fridman & Walheim, 2000; Schuck et al., 2004; Stokland et al., 2004). A thorough analysis was performed by Schuck et al. (2004), who identified seven general deadwood attributes. From them, Rondeux & Sanchez (2009) selected three basic parameters: tree species, dimensions (diameter, height or length), and decay stage. Other attributes, such as cause of death, presence of cavities, or amount of bark left, are less frequently used (Rondeux & Sanchez, 2009).

3.1 Parameters of deadwood

3.1.1 Tree species

Tree species is important particularly because it determines the decay rate of deadwood (Rondeux & Sanchez, 2009) and the habitat qualities for colonising species (Schuck et al., 2004). For recently dead trees, the identification of tree species is quite easy, but as decomposition advances, the identification becomes more difficult (Rondeux & Sanchez, 2009; Stokland et al., 2004). The best criteria are bark surface, and wood structure (Stokland et al., 2004). In Scandinavia, the angle between the trunk and the branches is also used as an identification criterion for separating between Pinus sylvestris and Picea abies, or between coniferous and broadleaved species (Stokland et al., 2004). If a tree species or a genus is impossible to determine, it is recommended to distinguish between conifers/broadleaves and/or hardwood/softwood (Rondeux & Sanchez, 2009; Schuck et al., 2004; Stokland et al., 2004), as many colonising species show preferences for a certain tree species or a group of species (Stokland et al., 2004). From 16 examined tree genera, Quercus was found to have the highest number of specialists (58 species, Jonsell et al., 1998). However, almost all tree genera have some monophagous species, i.e. species restricted to a single tree genus (Jonsell et al., 1998). According to Stokland et al. (2004), only 10% of the colonising species are generalists that utilise both coniferous and broadleaved tree species. As the wood decays, the host range broadens (Jonsell et al., 1998). In the latest stage of decomposition, when it is clearly impossible to determine tree species or a group, deadwood is classified as "unindentified" (Rondeux & Sanchez, 2009; Stachura et al., 2007; Stokland et al., 2004).

3.1.2 Dimensions

Dimensions are necessary attributes to calculate approximate volume of deadwood (Stokland et al., 2004). They are also used to distinguish between individual types of deadwood, e.g. coarse and fine woody debris, or snags and stumps. According to Stokland et al. (2004), diameter is one of the most important biodiversity attributes of deadwood, since a majority of species colonising deadwood respond to its value. While some species prefer small diameters up to 20 cm, other species colonise only deadwood with diameters greater than 20 cm or even 40 cm. Only about 20% of the examined species were found to be generalists (Stokland et al., 2004).

As already mentioned, the variability of minimum threshold diameter between coarse and fine woody detritus is large. In the context of forest management dead wood with diameter of more than 6-7 cm is generally accepted as "deadwood" (Atici et al. 2008), while smaller material is considered of lesser importance (Colak, 2002). Some authors (e.g. Debeljak, 2006; Travaglini et al. 2007; Vandekerkhove et al. 2009) use minimum threshold 5 or 7 cm, which equals the threshold diameter for living trees. Other authors (e.g. Atici et al., 2008; Fridman & Walheim, 2000; Nordén et al., 2004) use the diameter limit 10 cm as suggested by Harmon & Sexton (1996). However, Vandekerkhove et al. (2009) showed that the differences in diameter thresholds did not significantly affect the results, and concluded that unlike stand density, deadwood volume is less affected by size thresholds.

The main attributes used for describing standing deadwood are diameter and height (Rondeux & Sanchez, 2009). In case of standing dead trees and snags with height 1.3 m or more, diameter at breast height is measured. In case of stumps and snags shorter than 1.3 m, diameter at the level where the tree was cut or where the stem was broken off is measured (Travaglini et al., 2007). For lying deadwood, a diameter in the middle of the log (a so called mid diameter) and its length are the main attributes (Rondeux & Sanchez, 2009). Alternatively, diameters at the top and bottom ends of the log (i.e. top and basal diameter, respectively) can be measured in order to produce more precise volume calculations (Harmon & Sexton, 1996).

3.1.3 Decay stage

Decomposition is the process by which organic material is broken down into simpler forms of matter. If the decomposition is driven by physical and chemical processes, it is referred to as abiotic decomposition, while biotic decomposition is the degradation by living organisms, particularly microorganisms. It usually occurs in a number of sequential stages. As presented by Stokland et al. (2004), decay stage is an important quality influencing the associated species composition. In case of decay fungi, Veerkamp (2003) stated that it is the most important factor affecting their occurrence.

The decomposition of deadwood has been studied in many forests of the world (Bütler et al., 2007; Harmon et al., 2000; Jonsson, 2000; Krankina & Harmon, 1995; Kruys et al., 2002; Morelli et al., 2007; Næsset, 1999; Storaunet & Rolstand, 2002; Yatskov, 2001; etc.). The results of these studies showed that the rate/speed of decomposition of deadwood depends on a number of factors (Næsset, 1999), which can be divided into wood characteristics (tree species, dimensions) and site environmental factors (Radtke et al., 2004). Temperature and moisture seem to be the driving site parameters (Yin, 1999), but also the aspect of slope has an effect on the decomposition rate (Harmon et al., 1986). The contact of the deadwood with the ground is another important factor increasing the decay rate (Hytterborn & Packham, 1987, as cited in Næsset, 1999; Duvall & Grigal, 1999; Mattson et al., 1987).

Visual classification of deadwood is the most common approach of decomposition assessment (Bütler et al., 2007). The classification schemes generally distinguish several (3-7) decay stages (e.g. Holeksa, 2001; Lee et al., 1997; Maser et al., 1979; Næsset, 1999; Sollins, 1982). Each study defines its own class specification according to its objectives (Rondeux & Sanchez, 2009). The characterisation of decay stages/classes is usually based on the morphological features (log shape, wood texture, bark adherence, presence or absence of twigs and branches), hardness of the wood, and position with respect to the ground. The hardness of the wood, or the level of rottenness, is assessed by the depth a pushed knife penetrates the wood (Holeksa, 2001; Kuuluvainen et al., 2002; Rouvinen et al., 2005; Pasierbek et al., 2007). Other characteristics that have proven useful to distinguish decay classes include colour of wood, sloughing of wood, friability or crushability of wood (Harmon & Sexton, 1996). Some studies also use the presence of biological indicators (e.g. moss or plant cover, Holeksa, 2001), but Harmon & Sexton (1996) argue that they are of little value because they significantly vary even within a limited area.

Usually, the first decay class represents recently dead wood, which is least decayed with intact bark, present twigs and branches, round shape, smooth surface, intact texture, and the position elevated on support points. As the decay process proceeds the twigs, parts of branches and bark become traces to absent, wood becomes softer and fragmented, and the round shape becomes elliptical. The last decay class represents the most decomposed dead wood with no bark, twigs or branches, which is very soft, strongly fragmented and in contact with the ground along the whole length.

In order to quantify the decomposition process in a more objective way, it may be more straightforward to measure target variables. One method that involves ultrasonic measurements to characterize the wood quality of timber has been described by Sandoz (1989).

3.1.4 Volume, biomass, carbon

The volume of deadwood can either be calculated from the basic dimensions (diameter and height) or estimated on the base of the ocular judgement, which is the simplest and the most cost-effective, but a less accurate method. A thorough analysis of deadwood volume calculations is presented in Rondeux & Sanchez (2009). In general, the calculation of volume of individual standing and fallen dead trees follows the approaches applied to living trees (Rondeux & Sanchez, 2009). The most precise method is the volume equation derived for a particular species and region. Volume tables represent an easier, but a less precise approach (Šmelko, 2010).

In case of snags higher than 1.3 m, some authors (e.g. Travaglini & Chirici, 2006; Vacik et al., 2009) use the same approaches as for the living trees multiplied by a reduction factor representing the reduction of the tree height due to the top breakdown. Other studies approximate the volume of snags by common geometric solids (cylinders, cones, paraboloids), which the wood pieces resemble most (Rondeux & Sanchez, 2009). For example, Vandekerkhove et al. (2009) used the formulas of truncated cones. Merganičová & Merganič (2008) used an integral equation based on the models of stem shape derived by Petráš (1986, 1989, 1990).

Logs, i.e. lying deadwood pieces, usually resemble cylinders. There are three possibilities to calculate their volume depending on which diameters were measured. If only a mid diameter was measured, their volume is calculated using Huber´s formula. In case, top and bottom diameters are known, Smalian´s formula is used. Occasionally, all three diameters (top, middle, and bottom) are measured. In such cases, Newton´s formula can be applied. Concerning the question about the ideal formula to estimate log volume, Harmon & Sexton (1996) concluded that on average, all the formulae presented above should give satisfactory results. For individual logs, Newton's formula had the smallest average deviation from the "true" volume, but none of the formulae were biased (Harmon & Sexton, 1996). The volume of stumps is usually calculated using the formula of cylinder (e.g. Rouvinen et al., 2005).

Belowground volume of deadwood is usually omitted from the studies. Debeljak (2006) presented a method for the calculation of the belowground quantity of coarse woody debris based on the aboveground volume, ratio between the total volume to the root system volume, and the decay stage. Similarly, fine woody debris is only rarely included in the analyses.

Most of the above stated approaches dealt with the volume of individual deadwood pieces. To obtain a value representing the whole stand, the volumes of all occurring types of deadwood (dead trees, snag, logs, stumps) are summed up and converted to hectare values. The stand volumes of deadwood are used to compare different forests, regions, countries, and to evaluate the level of biodiversity, naturalness, or sustainability.

However, volume is not the best indicator if nutrient cycling and/or carbon sequestration are of primary interests, because during the decomposition process woody debris looses not only its volume, but also its mass and density (Coomes et al., 2002; Harmon et al., 2000; Krankina & Harmon, 1995). In such cases, biomass and carbon stock are estimated from volume. Carbon storage in wood is obtained by converting the volume mass into the amount of carbon stored in deadwood. For this conversion, carbon content in wood and wood density need to be known. Usually, carbon fraction in wood is approximated 50% of the woody dry mass (Coomes et al., 2002). Weiss et al. (2000) published more precise values for individual tree species of Central Europe. According to these authors, carbon content in Norway spruce wood is 50.1%, while in European beech 48.6%. This fraction remains stable during the whole decomposition process of deadwood (Bütler et al., 2007).

Unlike carbon fraction in wood, wood density decreases as wood decays (Harmon et al., 2000). While basic wood density of Norway spruce living trees fluctuates around 0.43 g cm^{-3} (Bütler et al., 2007; Morelli et al., 2007, Weiss et al., 2000), the average density of the most decayed Norway spruce deadwood is only 0.138 g cm^{-3} (Merganičová & Merganič, 2010). This is a significant reduction of wood density over the course of decomposition, which needs to be taken into account in the carbon stock studies. Merganičová & Merganič (2010) presented that when the volume of coarse woody debris was converted to carbon stocks using the basic wood density of fresh wood (i.e. 0.43 g cm^{-3}), deadwood carbon stocks were overestimated by 40% or more.

3.2 Deadwood inventory

Initially, information about deadwood was collected to address wildlife habitat issues (Bütler, 2003). Nowadays, deadwood is considered relevant for a number of different issues including nature conservation, forest certification, sustainable forest management, carbon sequestration, etc. Hence, deadwood is now assessed in many parts of the world within national forest inventories and various research activities. However, no harmonised methodology for deadwood inventory exists, as the objectives and the needs differ between the studies (Rondeux & Sanchez, 2009). In spite of these differences, simple, fast and accurate methods are required for both research and management purposes (Bütler, 2003).

In general, the survey can be accomplished either in the whole area of interest, or only on a portion of the given area. Although complete enumeration can provide us with the information about each individual, in large populations, such as forests, this survey is usually not economically and practically feasible. Hence, complete field inventory has been applied very scarcely, e.g. for the repeated measurements of virgin forests in the Czech republic (Vrška et al., 2001a, 2001b, 2001c), or to compare different assessment methods (e.g. Bütler, 2003). Because of high time and cost demands, sampling is often applied instead, while its main condition is that the selected sample represents the whole population. As Shiver & Borders (1996) pointed out, in most cases sampling can be more reliable than complete inventory, because more time can be taken to measurements, while sampling error could be kept small. There exist a great number of publications devoted to sampling techniques in general, e.g. Cochran (1977), Hush et al. (2003), Kish (1995), Shiver & Borders (1996), Schreuder et al. (1993), Šmelko (1985), Thompson (2002), Zöhrer (1980), etc., presenting different sampling designs from simple random sampling, through systematic and stratified sampling, up to multi-stage or multi-phase sampling designs. Deadwood sampling methods present specific characteristics with regard to the spatial distribution and the variability of deadwood components (Rondeux & Sanchez, 2009). As already mentioned, deadwood consists of different types (standing/lying, aboveground/belowground, coarse/fine woody debris) several of which would ideally require their own survey method. Due to this complexity, the majority of studies usually account only for the selected components of deadwood. For example, Marage & Lemperiere (2005) studied standing dead trees only. Most often the aboveground coarse woody debris is the subject of interest, while belowground deadwood is conventionally excluded because its assessment and quantification is difficult (Rondeux & Sanchez, 2009).

Standing deadwood is usually inventoried with the same methodology as living trees (Rondeux & Sanchez, 2009). The common approach is the plot-based sampling, while the plots can either be of fixed or variable area. An example of plots with variable area is when the plots are selected depending on a certain pre-defined number of trees to be included, e.g. the optimisation study of Šmelko (1968) proposed to measure 20-25 trees. In some cases, a set of concentric circles is used (Oehmichen, 2007; Šmelko & Merganič, 2008). Bütler (2003) and Vacik et al. (2009) applied the Bitterlich relascope point sampling, which uses a fixed angle of sight to select the trees to be assessed.

Lying deadwood can also be assessed on the same sample plots as living trees or standing deadwood. The only exception is the Bitterlich sampling method, which cannot be applied to downed deadwood directly (Vacik et al., 2009). In case of plot-based sampling, a common approach is to inventory only the deadwood inside the plot, i.e. if the log crosses the plot border, the part outside the plot is not accounted for in the inventory (Oehmichen, 2007). The second widely used approach is the line intersect sampling firstly presented by Warren & Olsen (1964) and Wagner (1968) for the inventory of logging waste and fuel wood, respectively. The principle of the method is that only the deadwood that crosses the line/transect is inventoried.

Both plot-based and line-based sampling techniques have some advantages and disadvantages. Line sampling is fast and accurate (Harmon & Sexton, 1996), easy to use, more time efficient and more economical than plot-based approach (Oehmichen, 2007). However, plot-based sampling is applicable to all types of deadwood, while line intersect method includes only lying deadwood. Hence, if this method is to be used for the deadwood assessment, it must be coupled with another method to consider standing deadwood. Additionally, Oehmichen (2007) and Rondeux & Sanchez (2009) stated that line intersect sampling is not suitable for long-term monitoring.

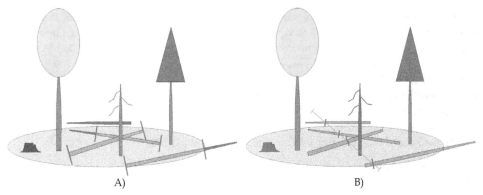

A)   B)

Fig. 3. Assessment of coarse woody debris using fixed-area (A) and line intersection (B) sampling techniques. Red colour indicates the elements of coarse woody debris that are not inventoried with the particular method.

Apart from the measurement-based inventory methods, the approach of visual estimation is also used, particularly for the inventory of fine woody debris, as its measurement would be too time consuming. With the visual assessment we can either directly estimate the volume of fine woody debris. However, more frequently other parameters, from which the volume is calculated, are estimated, e.g. average diameter and length (Rondeux & Sanchez, 2009), and the coverage of the plot by fine woody fractions (Šmelko et al., 2006; Šmelko & Merganič, 2008; Šmelko, 2010).

The methods presented so far are field inventory techniques. With the progress of modern technology, new approaches based on remote sensing are promising for coarse deadwood inventory. Infrared aerial photos are suitable for mapping and quantifying of standing deadwood (dead trees and snags) (Bütler, 2003; Bütler & Schlaepfer, 2004), but lying deadwood is not possible to map unless their resolution is 20 cm (Frei et al., 2003, as cited in Oehmichen, 2007). Pesonen et al. (2008) and Vehmas et al. (2009) used airborne laser scanning to assess coarse woody debris characteristics and stated that this approach can be used for the preliminary mapping of forests with large amounts of downed deadwood.

4. Functions of deadwood in forest ecosystems

4.1 Biotope

A great number of different plant and animal species have been associated with deadwood as a space for their lives or the parts of their lives. Over the last decades, the scientific studies have shown that deadwood provides valuable habitats for lichens, bryophytes, fungi, invertebrates, small vertebrates, birds, and mammals (Humphrey et al., 2004). Although the exact numbers differ between regions and studies, in general it has been assumed that around 25% of species occurring in forests are dependent on decaying wood (Schuck et al., 2004). The same number was reported by Siitonen (2001) and Stokland et al. (2003) for Scandinavia.

For saproxylic organisms, dead and dying trees and their parts are the key elements of their life (Davies et al., 2008; Grove, 2002; Lonsdale et al., 2008; Zhou et al., 2007). The group of saproxylic species is the most diverse group in forest ecosystems (Schuck et al., 2004), as it includes all species that depend on deadwood during some part of their life, or upon other saproxylics (Speight, 1989). The main saproxylic taxa are fungi (Ferris et al., 2000; Pouska et al., 2010; Siitonen, 2001), bryophytes (Kruys et al., 1999; Kushnevskaya et al., 2007; Odor & Standovar, 2001; Zielonka & Piatek, 2004), lichens (Kushnevskaya et al., 2007), arthropods such as beetles (Davies et al., 2008; Jonsell & Nordlander, 2002; Persiani et al., 2010), and birds (Bütler et al., 2004). From other groups with a smaller number of saproxylic species we can name mammals such as bats and dormice (Maser & Trappe, 1984), amphibians (DeMaynadier & Hunter, 1995), and moluscs (Kappes et al., 2009).

For other species, deadwood is a source of food and/or construction materials, a nesting and/or breeding site, a shelter, and a hiding place (Bütler, 2003; Debeljak, 2006). The tunnels left after wood-destroying insects can be used as a hiding place for other insects, e.g. wasps (Ehnstrom, 2001). Deadwood can also be used as a lookout post by mammals, e.g. squirrels (Bütler, 2003), or lynx (Bobiec et al., 2005).

4.2 Substrate

Coarse downed deadwood represents favourable environment for the natural regeneration of plant species starting from moss, ferns, and herbs (Franklin et al., 1987; Harmon et al., 1986; Radu, 2007). It is also a primary site to be colonised by fungi and an important seedbed for regenerating tree species. In some forests, the regeneration of tree species is exclusively dependent on the presence of deadwood (Nakagawa et al., 2001; Narukawa & Yamamoto, 2002; Narukawa et al., 2003; Takahashi et al., 2000). In some cases, tree seedlings occur only on the logs of the same tree species (Hofgaard, 1993), while in other cases they colonise the logs of other tree species (Harmon & Franklin, 1989). On deadwood, seedlings are provided with better temperature and moisture conditions (Mai, 1999). Due to these qualities, deadwood is an important rooting substrate particularly in cool climate and severe conditions of boreal (Harmon & Franklin, 1989) and mountain forests (Szewczyk & Szwagrzyk, 1991; Vorčák et al., 2005). Lepšová (2001) and Jaloviar et al. (2008) reported that the root system of the seedlings growing on woody debris was better developed. In ecosystems with abundant herb layer, the regeneration established on such "nurse logs" (Harmon et al., 1986) is favoured against the competing plants (Mai, 1999). However, not all coarse woody debris is a suitable place for regeneration. The suitability of downed deadwood for the regeneration depends on its qualitative parameters, mainly the decay stage, which is closely coupled to other characteristics (moisture, nutrient content). More decayed wood is more appropriate than fresh or slightly decayed deadwood (Merganič et al., 2003).

A)   B)

Fig. 4. Deadwood as biotope for saproxylic fungi (A) and substrate for new generation (B), photo by J. Vorčák.

4.3 Nutrient source

Deadwood significantly affects the flow of mass, energy and nutrients in ecosystems. The accumulation and decomposition of organic material on soil surface and in soil is closely coupled to nutrient cycling (Green et al., 1993). Although the relative concentration of nutrients in wood and bark is low, due to the large biomass amount of deadwood it is the main source of nutrients and carbon in forest ecosystems (Caza, 1993; Harmon et al., 1986). Nutrients are released from deadwood slowly over a long time; hence deadwood acts as a natural fertiliser (Holub et al., 2001; Hruška & Cienciala, 2002; Janisch & Harmon, 2001; Mackensen & Bauhus, 2003; Turner et al., 1995). Stevens (1997) reported that coarse woody debris is a nutrient source for more than 100 years. The release of nutrients can be performed by several different ways. Fungi and mosses growing on the surface take nutrients from deadwood, and hence, create a so-called zone of active nutrient cycling (Nadkarni & Matelson, 1992). The flow of nutrients runs through the mycelia of wood decomposing fungi (Zimmerman et al., 1995) as well as through the mycelia of ectomycorrhizal fungi (Lepšová, 2001). It is assumed that the colonisation of deadwood by fungi and microbes is one of the main phases of nutrient cycling (Caza, 1993). The microorganisms and wood-decomposing fungi decompose organic molecules of wood, and thus release the nutrients for plants (Stevens, 1997). All ways of nutrient release are directly or indirectly coupled to the processes causing wood decay. In addition, deadwood is also an important resource of water, especially in dry periods (Harmon & Sexton, 1995; Pichler et al., 2011).

5. Importance of deadwood

5.1 Productivity

The presence of deadwood in forest ecosystems increases their productivity (Marra & Edmonds, 1994; Mcminn & Crossley, 1993). Debeljak (2006) compared managed and virgin forest stands and found that the maximum tree height was significantly lower in the managed stands which may indicate the reduction of forest productivity due to the reduction of coarse woody debris.

5.2 Biodiversity

Since deadwood has been recognised as a habitat of great importance for many species of forest ecosystems, it is considered to be a key element of biodiversity in forests. In many cases, deadwood is associated with relict, rare and protected species (Radu, 2007), and therefore, it is regarded as a key feature for the preservation of many threatened species (Ranius et al., 2003). The recognition of the deadwood importance for biodiversity has led to the incorporation of its quantitative parameters in biodiversity monitoring programmes (Humphrey et al., 2004), e.g. the European Environmental Agency includes deadwood as one of 15 core biodiversity indicators (Kristensen, 2003). The results of the different studies suggest, that the importance of deadwood and its individual parameters differs between the sites and cannot be generalised. According to Humphrey et al. (2004), coarse woody debris is an important biodiversity indicator in conifer-dominated forests in the Atlantic and Boreal biogeographical regions, but is less applicable to Mediterranean forests.

The biological importance of deadwood depends on several factors, in particular on tree species, dimension, vertical and horizontal position, decay stage, and micro-environmental conditions (Radu, 2007). From all deadwood characteristics, the amount of deadwood is usually taken as an indicator of biodiversity (Stokland et al., 2004; Vandekerkhove et al., 2009). A higher amount of deadwood in forests increases the number, and the density of species, and hence, species richness, because higher deadwood amount means greater surface and area in forests, and hence, its higher availability for potential users (Müller & Bütler, 2010). This is in accordance with the island theory (Cook et al., 2002) according to which we can expect higher number of species in a unit with a larger "island". Secondly, larger surface means greater possibilities of its differentiation (Müller & Bütler, 2010) providing species with a greater variability of habitats.

Several studies detected the correlation of species richness to deadwood amount (Müller & Bussler, 2008; Müller & Bütler, 2010). It is widely agreed that 20-30 $m^3 \cdot ha^{-1}$ is the required amount that can safeguard the complete spectrum of deadwood-depending species (Angelstam et al., 2003; Humphrey et al., 2004; Siitonen, 2001; Vandekerkhove et al., 2009), although some authors presented higher values. For example, according to Müller & Bussler (2008), the critical threshold for saproxylic Coleoptera is between 40 and 60 $m^3 \cdot ha^{-1}$. Other studies argued that for some taxa the diversity of deadwood components, and its continuity in space and time is a more important feature than its amount (Heilmann-Clausen & Christensen, 2004; Schiegg, 2000a; Similä et al. 2003). Hence, for biodiversity conservation it is also important to balance the proportions of individual deadwood types and decay stages (Hagan & Grove, 1999), as each type represents quite different habitats suitable for different species.

5.3 Naturalness

Deadwood characteristics are also common attributes used for assessing forest naturalness (Laarmann et al., 2009; Winter et al., 2010), since the differences in deadwood between managed and unmanaged forests are in most cases notable (Kirby et al., 1998; Liira et al., 2007). In general, the amount of deadwood is lower in managed in stands than the stands left to self-development (Rahman et al., 2008; Rondeux & Sanchez, 2009; Travaglini & Chirici, 2006), because most of large-sized wood is extracted from the forest. Therefore, to determine the level of naturalness of forest ecosystems, the amount of coarse woody debris has become an important indicator (Rahman et al., 2008). However, for the proper application of deadwood as an indicator, reference values representing the "natural" state are needed. Such values can be obtained from natural unmanaged forests (virgin/old-growth forests) (Hahn & Christensen, 2004; Humphrey et al., 2004). However, when using the reference values it must be considered that the volume of deadwood varies considerably among forest sites. While in North America the volume of coarse woody debris in old-growth forests may exceed 1,000 $m^3 \cdot ha^{-1}$ (Harmon et al., 1986), in Europe the average values between 40 and 200 $m^3 \cdot ha^{-1}$ have been reported (Albrecht, 1991; Christensen et al., 2005; Hort & Vrška, 1999; Karjalainen et al., 2002; Siitonen et al., 2000; Vallauri et al., 2003; Vandekerkhove et al., 2009). Larger values (more than 400 $m^3 \cdot ha^{-1}$) have been found in virgin forests of Central Europe, e.g. Slovakia (Saniga and Schütz, 2001), Poland (Bobiec, 2002), Slovenia (Debeljak, 1999, 2006). Even in Europe, there is a large variability of deadwood accumulation, when northern boreal and southern Mediterranean forests are characterised by a lower deadwood amount than Central European mixed forest types (Hahn & Christensen, 2004).

Large amount of deadwood does not necessarily indicate the high level of naturalness of forest ecosystems (Laarman et al., 2009; Marage & Lemperiere, 2005; Rouvinen et al., 2005; Pasierbek et al., 2007), particularly in cases when it is the result of the accelerated breakdown of forest stands (Jankovský et al., 2004) due to disturbances. The analysis of Laarman et al. (2009) reported that the spatial distribution of deadwood, proportion of recent mortality and the causes of mortality seem to be better indicators. Marage & Lemperiere (2005) revealed that the differences between the managed and unmanaged forest stands were only in the category of standing dead trees, as in the managed stands the trees are extracted before they reach their maximum diameter. Hence, deadwood is an appropriate indicator of naturalness only if the additional information about its components, decay stages, etc. is also available (Liira & Sepp, 2009; Rondeux & Sanchez, 2009).

5.4 Geomorphology

Mechanical and physical qualities of large-sized deadwood have a significant influence on the geomorphology of forest soils and small watercourses in forest ecosystems (Stevens, 1997). On slopes, coarse woody debris significantly contributes to slope and soil surface stability. Fallen logs prevent or slow down soil erosion and surface water runoff (Kraigher et al., 2002; Stevens, 1997), and act as barriers for rock fall and avalanches (Kupferschmidt et al., 2003).

6. Factors influencing deadwood pool in forest ecosystems

Deadwood is the result of the influence of various abiotic and biotic factors on individual trees or forest stands. The quantity of deadwood and its decomposition in a particular forest ecosystem depends on many intrinsic and extrinsic factors that drive the input of deadwood and its decomposition process. Intrinsic factors include deadwood type, dimensions, and tree genus that determines basic tree and wood characteristics, while extrinsic factors include climate and site conditions, and disturbances.

Tree species can predetermine the mortality pattern and subsequently the accumulation and decomposition of deadwood. Species with a shallow root system, e.g. Norway spruce, are susceptible to uprooting, while deep-rooted species usually suffer from breakdown. However, under favourable climate and soil conditions, this may not hold, as e.g. in mountain spruce forests of Babia hora Holeksa (2001) detected that the snags were dominant over the uprooted trees, whose proportion was much lower than in other temperate and boreal spruce forests.

The rate of deadwood decay depends on the chemical composition of wood, which is specific for each tree species. Some tree species are decay resistant, for example oak or pine (Radu, 2007). Softwood species, e.g. willow, birch, and poplar, have a much shorter period of decomposition (Radu, 2007). The resistance of deadwood depends on the content of extractives (polyphenols, waxes, oils, resins, gums, tannins) in the heartwood, which are toxic to most decay fungi and some insects, and hence slower the decomposition process.

However, the decay resistance of a particular species may vary greatly depending on the dimensions and site conditions. The length of the decomposition process is positively correlated with deadwood diameter. The turnover of fine dead woody debris is fast (Stevens, 1997), while its decomposition rate increases exponentially with decreasing diameter (Harmon & Sexton, 1996). The differences in decay rate are also between standing and lying deadwood. The decay of snags is slower than the decay of logs because of lower wood moisture (Kupferschmidt et al., 2003). Snag diameter has also a positive influence on the time length the snag stands (Everett et al., 1999; Morrison & Raphael, 1993, as cited in Bütler, 2003).

From site conditions, temperature and moisture seem to be the driving factors. In warm, moist environments decay rates are higher, because such conditions favour microbial and fungal growth (Yin, 1999). This is evident in the analysis of deadwood volume along the elevation gradient. According to MCPFE (2007), the lowest amount of deadwood is in the forests situated at the lowest elevations. As elevation increases, accumulated deadwood volume enlarges. This increasing trend with elevation was also observed in the national analysis in Slovakia (Merganič et al., 2011), when at the lowest elevations the authors reported only 10 m³·ha⁻¹, while under the timber line the deadwood volume was around 100 m³·ha⁻¹. Similarly, Kühnel (1999) reported that in mountain forests the amount of deadwood is three times higher than in lowlands.

6.1 Disturbances

In natural forests deadwood originates from tree mortality, which is either the result of inter-tree competition or senescence processes, or it is caused by natural disturbances, which can differ in terms of quality and quantity (Rahman et al., 2008). Disturbances can be driven by abiotic (wind, fire) and/or biotic factors (insect outbreak). Windthrow, icebreak, insect and fungal attacks leave various amounts of deadwood in the forest, while during fire disturbances deadwood is immediately consumed (Hahn & Christensen, 2004). The disturbance factors vary in intensity and scale leading to a patchy distribution of deadwood at the stand and landscape levels (Humphrey et al., 2004). Small-scale disturbances cause the death of individual trees or small groups of trees, while large-scale disturbances affect the whole ecosystem (Korpeľ, 1995).

A)   B)

Fig. 5. Large-scale disturbances caused by windthrow (A) and insect outbreak (B).

Small-scale events occur frequently and hence provide a continuous supply of deadwood (Rahman et al., 2008). Due to this, in forests following a so-called small cycle deadwood of different size and decay stages can be found (Korpeľ, 1995; Saniga & Schütz, 2001). On the landscape level, the deadwood pool is relatively stable, and the differences are obvious only on a small scale between the developmental stages. The highest amount of deadwood occurs in breakdown developmental stages, while the lowest amount is in the stage of maturity (Merganičová et al., 2004; Merganičová & Merganič, 2010).

In contrast, large-scale disturbances cause abrupt changes of the whole ecosystem, which result in high deadwood inputs at the time of the event. At this stage, these ecosystems may attain higher amounts of deadwood than the forests developing in a small cycle (Rahman et al., 2008; Šebeň et al., 2009). However, in the following successional stages, deadwood input is minimal, and the total amount of deadwood declines (Rahman et al., 2008).

6.2 Forest management

Apart from the natural factors discussed so far, a man with its activities also affects deadwood in forest ecosystems either directly through forest management or indirectly by e.g. air pollution (Debeljak, 2006). While the first forms of human utilisation of wood, e.g. domestic use of wood, had apparently only a slight influence (Rouvinen et al., 2005), with the increasing human population, human influence has become more intensive and widespread (Björn, 1999, as cited in Rouvinen et al., 2005). Nowadays, it is generally agreed that forest management has a negative effect on deadwood amount and deadwood components (Atici et al., 2008). Dead, damaged and weakened trees are often removed from the forest during harvesting operations (Bütler, 2003). Thinnings reduce deadwood inputs from natural mortality. Short rotation time limits the presence of large dead trees (Bütler, 2003; Debeljak, 2006; Marage & Lemperiere, 2005), because trees are harvested before they reach their maximum diameter (Atici et al., 2008). After natural disturbances, such as windthrow or insect outbreak, dead trees are harvested in salvage logging (Bütler, 2003). The effect of management on deadwood is linked with the accessibility of harvesting areas. Bütler (2003) found a significant negative relationship between road density and deadwood amounts in Switzerland. Pasierbek et al. (2007) detected more deadwood on the sites which were situated far from the settlements.

The current amount of aboveground deadwood in managed forest is very low. According to Nilsson et al. (2002), before human exploitation common deadwood amount in European forests was 130-150 $m^3 \cdot ha^{-1}$, nowadays the values range from 1 to 23 $m^3 \cdot ha^{-1}$ (MCPFE, 2007). Usually, the average amount of aboveground deadwood does not exceed 10 $m^3 \cdot ha^{-1}$ (FAO, 2000; Fridman & Walheim, 2000; Christensen et al., 2005). In some cases, it does not even reach 5 $m^3 \cdot ha^{-1}$ (Albrecht, 1991; Smykala, 1992; Schmitt, 1992; Vallauri et al., 2009). In addition, deadwood in managed forests typically consists of fine woody debris (small twigs and branches) and short stumps (Atici et al., 2008). A large, but often a forgotten part of woody debris in managed stands may be belowground deadwood originating from root systems of cut trees (Debeljak, 2006). This author presented that the proportion of belowground coarse woody debris in managed forests is significantly higher than in virgin forests. However, this type of deadwood is buried, and therefore not available to all saproxylic species, e.g. birds.

Therefore, in the interests of sustainable forestry and biodiversity conservation the efforts are being made to increase the levels of deadwood in managed forests (Atici et al., 2008). There exist a great number of publications that deal with the question how much deadwood should be left in a forest. However, the variation of the recommended values is quite large. Older studies suggest at least 3 $m^3 \cdot ha^{-1}$ (Utschik, 1991) or 5-10 $m^3 \cdot ha^{-1}$ (Ammer, 1991), which is 1-2% of the total volume of the stand. In more recent studies the suggested values are higher and fluctuate between 15 and 30 $m^3 \cdot ha^{-1}$ (Bütler & Schlaepfer, 2004; Colak, 2002; Jankovský et al., 2004), or 5-10% of the total stand volume (Bütler & Schlaepfer, 2004;

Jedicke, 1995; Möller, 1994; Vandekerkhove et al., 2009). According to Vandekerkhove et al. (2009), the minimum amount of deadwood should secure the existence of the whole spectrum of saproxylic species. The works dealing with their populations suggest 40 m³·ha⁻¹ to be the threshold value, when the diversity of saproxylic communities is comparable with the diversity in virgin forests (Haase et al., 1998; Kirby et al., 1998; Müller & Bussler, 2008). It is clear that very small values are too low to be important for nature conservation (Scherzinger, 1996) or biodiversity. Considering the large variation in the values of deadwood from natural forests, which are references for management, no universal value valid throughout the world exists (Jankovský et al., 2004). The guidelines and recommendations should be given for more homogeneous groups, e.g. individual forest types (Hahn & Christensen, 2004). The consensus should also be found between gains and losses (Atici et al., 2008).

7. Conclusion

Nowadays it is generally accepted that deadwood plays an important role in ecosystem dynamics. In this chapter we presented the synthesis of existing knowledge about deadwood including its assessment, evaluation, and factors influencing deadwood occurrence and dynamics. Although less than one decade ago, the information about the decay rates of deadwood in Europe was almost completely missing, this gap is currently being filled. Addressing the full spectrum of processes within forest ecosystems, in which deadwood occurs, we implied to coherently analyse its importance for sustainable forest management and issues closely coupled, e.g. biodiversity, naturalness. While the importance of deadwood for biodiversity has been thoroughly studied, less information is available about the relationship between deadwood and stand productivity and geomorphology. Future research should focus more on investigating belowground component of deadwood, which is currently usually not accounted for in the inventories. For forest management, more specific guidelines with regard to forest type and/or region would be helpful to maintain sufficient amount of deadwood for biodiversity preservation.

8. Acknowledgment

This work was supported by the National Agency for Agriculture Research in the Czech Republic under the contract No. QH91077, by the Ministry of Agriculture of the Czech Republic under the contract No. QI QI102A085 and the Slovak Research and Development Agency under the contracts No. APVT-27-009304, and APVV-0632-07.

9. References

Albrecht, L. (1991). Die Bedeutung des toten Holzes im Wald. *Forstwissenschaftliches Centrablatt*, Vol. 110, No. 2, pp. 106-113

Ammer, U. (1991). Konsequenzen aus den Ergebnissen der Totholzforschung für die forstliche Praxis. *Forstwissenschaftliches Centrablatt*, Vol. 110, pp. 149-157

Angelstam, P.; Breuss, M. & Gossow, H. (2000). Visions and tools for biodiversity restoration in coniferous forest. *Forest-ecosystem-restoration:-ecological-and-economical-impacts-of-restoration-processes-in-secondary-coniferous-forests-Proceedings-of-the-International-Conference,*-Vienna,-Austria,-10-12-April, pp. 19-28

Angelstam, P.K.; Butler, R.; Lazdinis, M.; Mikusinski, G. & Roberge, J.M. (2003). Habitat thresholds for focal species at multiple scales and forest biodiversity conservation—dead wood as an example. *Annales Zoologici Fennici*, Vol. 40, pp. 473–484

Atici, E.; Colak, A.H. & Rotherham, I.D. (2008). Coarse Dead Wood Volume of Managed Oriental Beech (Fagus orientalis Lipsky) Stands in Turkey. *Investigación Agraria: Sistemas y Recursos Forestales*, Vol. 17, No. 3, pp. 216-227 Available from http://biblion.epfl.ch/EPFL/theses/2003/2761/EPFL_TH2761.pdf

Bobiec, A. (2002). Living stands and dead wood in the Bialowieza forest: suggestions for restoration management. *Forest Ecology and Management*, Vol. 165, pp. 125-140

Bobiec, A.; Gutowski, J.M.; Laudenslayer, W.F.; Pawlaczyk, P. & Zub, K. (2005). *The Afterlife of a Tree*. Warsaw. WWF Poland

Butler, J.; Alexander, K.N.A. & Green, T. (2002). *Decaying wood: an overview of its status and ecology in the United Kingdom and Continental Europe*. USDA Forest Service Gen. Tech. Rep.PSW-GTR-181, pp. 11-19

Butler, R. & Schlaepfer, R. (2004). Dead wood in managed forests: how much is enough? *Schweizerische Zeitschrift für Forstwesen*, Vol. 155, No. 2, pp. 31-37

Bütler, R.; Angelstam, P.; Ekelund, P. & Schlaeffer, R. (2004). Dead wood threshold values for the three-toed woodpecker presence in boreal and sub-Alpine forest. *Biological Conservation*, Vol. 119, No. 3, pp. 305-318

Bütler, R.; Patty, L.; Le Bayon, R.C.; Guenat, C. & Schlaepfer, R. (2007). Log decay of Picea abies in the Swiss Jura Mountains of central Europe. *Forest Ecology and Management*, Vol. 242, pp. 791-799

Bütler, S.R. (2003). *Dead wood in managed forests: how much and how much is enough? Development of a Snag Quantification Method by Remote Sensing & GIS and Snag Targets Based on Three-toed Woodpeckers' Habitat Requirements*. PhD. Thesis, Lausanne EPFL 184 p., 20.01.2010, Available from http://biblion.epfl.ch/EPFL/theses/2003/2761/EPFL_TH2761.pdf

Caza, C.L. (1993). *Woody debris in the forests of British Columbia: a review of the literature and current research*. B.C. Ministry of Forests Land Management Report No. 78. Victoria, B.C.: 99

Christensen, M.; Hahn, K.; Mountford, E.P.; Odor, P.; Standovar, T.; Rozenbergar, D.; Diaci, J.; Wijdeven, S.; Meyer, P.; Winter, S. & Vrska, T. (2005). Dead wood in European beech (Fagus sylvatica) forest reserves. *Forest Ecology and Management*, Vol. 210, pp. 267-282

Cienciala, E.; Tomppo, E.; Snorrason, A.; Broadmeadow, M.; Colin, A.; Dunger, K.; Exnerova, Z.; Lasserre, B.; Petersson, H.; Priwitzer, T.; Sanchez, G. & Ståhl, G. (2008). Preparing emission reporting from forests: Use of National Forest Inventories in European countries. *Silva Fennica*, Vol. 42, No. 1, pp. 73-88

Cochran, W.G. (1977). *Sampling Techniques*. John Wiley 8 Sons, Inc., 428 p.

Colak, A.H., (2002). Dead wood and its role in nature conservation and forestry: a Turkish perspective. *The Journal of Practical Ecology and Conservation*, Vol. 5, No. 1, pp. 37-49

Cook, W.M.; Lane, K.T.; Foster, B.L. & Holt, R.D. (2002). Island theory, matrix effects and species richness patterns in habitat fragments. *Ecology Letters*, Vol. 5, No. 5, pp. 619-623

Coomes, D.A.; Allen, R.B.; Sćoty, N.A.; Goulding, C. & Beets, P. (2002). Designing systems to monitor carbon stocks in forests and shrublands. *Forest Ecology and Management*, Vol. 164, pp. 89-108

Davies, Z.G.; Tyler, C.; Stewart, G.B. & Pullin, A.S. (2008). Are current management recommendations for conserving saproxylic invertebrates effective? *Biodiversity Conservation*, Vol. 17, pp. 209-234

Debeljak, M. (1999). Dead trees in the virgin forest of Pecka. *Forestry Wood Sci. Technol.*, Vol. 59, pp. 5-31

Debeljak, M. (2006). Coarse woody debris in virgin and managed forest. *Ecological Indicators*, Vol. 6, pp. 733-742

DeMaynadier, P.G. & Hunter, M.L. (1995). The relationship between forest management and amphibian ecology: a review of the North American literature. *Environmental Review*, Vol. 3, pp. 230-261

Duvall, M.D. & Grigal, D.F. (1999). Effects of timber harvesting on coarse woody debris in red pine forests across the Great Lakes states, U.S.A. *Canadian Journal of Forest Research*, Vol. 29, pp. 1926-1934

Ehnstrom, B. (2001). Leaving dead wood for insects in boreal forests—suggestions for the future. *Scandinavian Journal of Forest Research Supplement*, Vol. 3, pp. 91-98

Elton, C.S. (1966). *Dying and dead wood*, In: The patterns of animal communities. New York: John Wiley and Sons, Inc.; pp. 279-305

FAO (2000). *Global Forest Resources Assessment 2000*, Rome, Italy.

Feller, M.C. (2003). Coarse woody debris in the old-growth forests of British Columbia. *Environmental Review*, Vol. 11, pp. 135-157

Ferris, R. & Humphrey, J.W. (1999). A review of potential biodiversity indicators for application in British forests. *Forestry-Oxford*, Vol. 72, No. 4, pp. 313-328

Ferris, R.; Peace, A.J. & Newton, A.C. (2000). Macrofungal communities of lowland Scots pine (Pinus sylvestris L.) and Norway Spruce (Picea abies (L.) Karsten) plantations in England: Relationships with site factors and stand structure. *Forest Ecology and Management*, Vol. 131, pp. 255-267

Franklin, J.F.; Shugart, H.H. & Harmon, M.E. (1987). Tree death as an ecological process. *BioScience*, Vol. 37, No .8, pp. 550-556

Fridman, J. & Walheim, M. (2000). Amount, structure, and dynamics of dead wood on managed forestland in Sweden. *Forest Ecology and Management*, Vol. 131, pp. 23-36

Green, R.N.; Trowbridge, R.L. & Klinka, K. (1993). *Towards a taxonomic classification of humus forms*. Forest Science, 39, Monograph, Vol. 29, 49 p.

Grove, S.J. (2002). Saproxylic insect ecology and the sustainable management of forests. *Annual Review of Ecology and Systematics*, Vol. 33, pp. 1-23

Haase, V.; Topp, W. & Zach, P. (1998). Eichen-Totholz im Wirtschaftswald als Lebensraum fur xylobionte Insekten, *Z. Ökologie u. Naturschutz*, Vol. 7, pp. 137-153

Hagan, J.M. & Grove, S.L. (1999). Coarse woody debris. *Journal of Forestry*, January, pp. 6-11

Hahn, K.; Christensen, M. (2004). Dead wood in European forest reserves - a reference for forest management. *EFI-Proceedings*, Vol. 51, pp. 181-191

Harmon, M.E. & Franklin, J.F. (1989). Tree seedlings on logs in Picea-Tsuga forests of Oregon and Washington. *Ecology*, Vol. 70, pp. 48-49

Harmon, M.E. & Sexton, J. (1996). *Guidelines for measurements of woody detritus in forest ecosystems*. U.S.LTER Network Office, University of Washington, Seatlle, Publ. No. 20.

Harmon, M.E.; Franklin, J.F.; Swanson, F.J.; Sollins, P.; Gregory, S.V.; Lattin, J.D.; Anderson, N.H.; Cline, S.P.; Aumen, N.G.; Sedell, J.R.; Lienkaemper, G.W.; Cromack, K. & Cummins, K.W. (1986): Ecology of coarse woody debris in temperate ecosystems. *Advances in Ecological Research*, Vol. 15, pp. 133-302

Harmon, M.E.; Krankina, O.N. & Sexton, J. (2000). Decomposition vectors: a new approach to estimating woody detritus decomposition dynamics. *Canadian Journal of Forest Research*, Vol. 30, pp. 76-84

Heilmann-Clausen, J. & Christensen, M. (2004). Does size matter? On the importance of various dead wood fractions for fungal diversity in Danish beech forests. *Forest Ecology and Management*, Vol. 201, pp. 105-117

Hofgaard, A. (1993). Structure and regeneration patterns in a virgin Picea abies forest in northern Sweden. *Journal of Vegetation Science*, Vol. 4, pp. 601-608

Holeksa, J. (2001). Coarse woody debris in a Carpathian subalpine spruce forest. *Forstwissenschaftliches Centralblatt*, Vol. 120, pp. 256-270

Holub, S.M.; Lajtha, K. & Spears, J.D. (2001). A reanalysis of nutrient dynamics in coniferous coarse woody debris. *Canadian Journal of Forest Research*, Vol. 31, pp. 1894-1902

Hort, L. & Vrška, T., 1999. Podíl odumřelého dřeva v pralesovitých rezervacích ČR, In: *Význam a funkce odumřelého dřeva v lesních porostech*. Vrška, T. (Ed.), Sbor. ref. NP Podyjí, Vranov nad Dyjí, pp. 75-86

Hruška, J. & Cienciala, E. (2002). Dlouhodobá acidifikace a nutriční degradace lesních půd – limitující faktor současného lesnictví. *MŽP*: 160

Humphrey, J.W.; Sippola, A.L.; Lemperiere, G.; Dodelin, B.; Alexander, K.N.A. & Butler, J.E. (2004). Deadwood as an indicator of biodiversity in European forests: from theory to operational guidance. *EFI-Proceedings*, Vol. 51, pp. 193-206

Husch, B.; Beers, T.W. & Kershaw, J.A.Jr. (2003). *Forest mensuration*. John Wiley 8 Sons, Inc., 443 p.

IPCC (2003). *Good practice guidance for land use, land-use change and forestry*. In: Penman, J.; Gytarsky, M.; Hiraishi, T.; Krug, T.; Kruger, D.; Pipatti, R.; Buendia, L.; Miwa, K.; Ngara, T.; Tanabe, K. & Wagner, F. (Ed.), IPCC/OECD/IEA/IGES, Hayama, Japan. 20.01.2010, Available from http:// www.ipcc-nggip.iges.or.jp/public/gpglulucf/gpglulucf_contents.htm

Jaloviar, P.; Szeghö, P. & Kucbel, S. (2008). The influence of coarse woody debris decomposition degree on selected fi ne roots' parameters of Norway spruce natural regeneration in NNR Babia hora. *Beskydy*, Vol. 1, No. 2, pp. 135-142

Janisch, J.E. & Harmon, M.E. (2001). Successional changes in live and dead wood carbon stores: implications for net ecosystem productivity. *Tree Physiology*, Vol. 22, pp. 77-89

Jankovský, L.; Lička, D. & Ježek, K. (2004). Inventory of dead wood in the Kněhyně-Čertův mlýn National Nature Reserve, the Moravian-Silesian Beskids. *Journal of Forest Science*, Vol. 50, No. 4, pp. 171-180

Jedicke, E., (1995). Anregungen zu einer Neuauflage des Altholzinsel- Programms in Hessen. *Allgemeine Forstzeitung*, Vol. 10, pp. 522-524

Jonsell, M. & Nordlander, G. (2002). Insects in polypore fungi as indicator species: a comparison between forest sites differing in amounts and continuity of dead wood. *Forest Ecology and Management*, Vol. 157, pp. 101-118

Jonsell, M.; Weslien, J. & Ehnström, B. (1998). Substrate requirements of red-listed saproxylic invertebrates in Sweden. *Biodiversity and Conservation*, Vol. 7, pp. 749-764

Jonsson, B.G. (2000). Availability of coarse woody debris in a boreal old-growth Picea abies forest. *Journal of Vegetation Science*, Vol. 11, pp. 51-56

Kappes, H.; Jabin, M.; Kulfan, J.; Zach, P. & Topp, W. (2009). Spatial patterns of litter-dwelling taxa in relation to the amount of coarse woody debris in European temperate deciduous forests. *Forest Ecology and Management*, Vol. 257, pp. 1255-1260

Karjalainen, L.; Kuuluvainen, T. & Korpilahti, E. (2002). Amount and diversity of coarse woody debris within a boreal forest landscape dominated by Pinus sylvestris in Vienansalo wilderness, eastern Fennoscandia. *Silva Fennica*, Vol. 36, pp. 147-167

Kirby, K.J.; Reid, C.M.; Thomas, R.C. & Goldsmith, F.B. (1998). Preliminary estimates of fallen dead wood and standing dead trees in managed and unmanaged forests in Britain. *Journal of Applied Ecology*, Vol. 35, pp. 148-155

Kish, L. (1995). *Survey sampling*. John Wiley 8 Sons, Inc., 643 p.

Korpeľ, Š. (1989). *Pralesy Slovenska*. Veda, SAV, Bratislava, 332 p.

Kraigher, H.; Jurc, D.; Kalan, P.; Kutnar, L.; Levanic, T.; Rupel, M. & Smolej, I. (2002). Beech coarse woody debris characteristics in two virgin forest reserves in southern Slovenia. *Forestry Wood Science Technology*, Vol. 69, pp. 91–134

Krankina, O.N. & Harmon, M.E. (1995). Dynamics of the dead wood carbon pool in Northwestern Russian boreal forests. *Water, Air and Soil Pollution*, Vol. 82, pp. 227-238

Krankina, O.N.; Harmon, M.E. & Griazkin, A.V. (1999). Nutrient stores and dynamics of woody detritus in a boreal forest: modeling potential implications at the stand level. *Canadian Journal of Forest Research*, Vol. 29, pp. 20-32

Kristensen, P. (2003). *EEA core set of indicator*. Revised version April, 2003. Copenhagen, Denmark: European Environment Agency, 24.08.2011, Available from http://www.unece.org/env/europe/monitoring/StPetersburg/EEA%20Core%20 Set%20of%20Indicators%20rev2EECCA.pdf

Kruys, N.; Fries, C.; Jonsson, B.G.; Lamas, T. & Stahl, G. (1999). Wood-inhabiting cryptogams on dead Norway spruce (Picea abies) trees in managed Swedish boreal forests. *Canadian Journal of Forest Research*, Vol. 29, pp. 178-186

Kruys, N.; Jonsson, B.G. & Stahl, G. (2002). A stage-based matrix model for decay-class dynamics of woody debris. *Ecological Applications*, Vol. 12, No. 3, pp. 773-781

Kupferschmidt, A.D.; Brang, P.; Schönenberger, W. & Bugmann, H. (2003). Decay of Picea abies snags stands on steep mountain slopes. *The Forestry Chronicle*, Vol. 79, No. 2, pp. 1-6

Kushnevskaya, H.; Mirin, D. & Shorohova, E. (2007). Patterns of epixylic vegetation on spruce logs in late-successional boreal forests. *Forest Ecology and Management*, Vol. 250, pp. 25-33

Kuuluvainen, T.; Aapala, K.; Ahlroth, P.; Kuusinen, M.; Lindholm, T.; Sallantaus, T.; Siitonen, J. & Tukia, H. (2002). Principles of ecological restoration of boreal forested ecosystems: Finland as an example. *Silva Fennica*, Vol. 36, pp. 409–422

Laarmann, D.; Korjus, H.; Sims, A.; Stanturf, J.A.; Kiviste, A. & Koester, K. (2009). Analysis of forest naturalness and tree mortality patterns in Estonia. *Forest Ecology and Management*, Vol. 258S, 5187-5195

Lee, P.C.; Crites, S.; Nietfeld, M.; Nguyen, H.V. & Stelfox, J.B. (1997). Characteristics and origins of deadwood material in aspen-dominated boreal forests. *Ecological Applications*, Vol. 7, pp. 691-701

Lepšová, A. (2001). Ectomycorrhizal system of naturally established Norway spruce [Picea abies (L.) Karst] seedlings from different microhabitats – forest floor and coarse woody debris. *Silva Gabreta*, Vol. 7, pp. 223-234

Lexer, M.J.; Lexer, W. & Hasenauer, H. (2000). The Use of Forest Models for Biodiversity Assessments at the Stand Level. *Invest. Agr.: Sist. Recur. For.: Fuera de Serie n.º 1*, pp. 297-316

Liira, J. & Sepp, T. (2009). Indicators of structural and habitat quality in boreo-nemoral forests along the management gradient. *Ann. Bot. Fennici*, Vol. 46, pp. 308-325

Liira, J.; Sepp, T. & Parrest, O. (2007). The forest structure and ecosystem quality in conditions of anthropogenic disturbance along productivity gradient. *Forest Ecology and Management*, Vol. 250, pp. 34-46

Lonsdale, D.; Pautasso, M. & Holdenrieder, O. (2008). Wood-decaying fungi in the forest: conservation needs and management options. *European Journal of Forest Research*, Vol. 127, pp. 1-22

Mackensen, J. & Bauhus, J. (2003). Density loss and respiration rates in coarse woody debris of Pinus radiata, Eucalyptus regnans and Eucalyptus maculata. *Soil Biology and Biochemistry*, Vol. 35, pp. 177-186

Mai, W. (1999). Über Ammenstäemme im Gebirgswald. *LWF aktuell*, Vol. 18, pp. 18-20

Marage, D. & Lemperiere, G. (2005). The management of snags: A comparison in managed and unmanaged ancient forests of the Southern French Alps. *Annals of Forest Science*, Vol. 62, No. 2, pp. 135-142

Marra, J.L. & Edmonds, R.L. (1994). Coarse woody debris and forest floor respiration in an old-growth coniferous forest on the Olympic Peninsula, Washington, USA. *Canadian Journal of Forest Research*, Vol. 24, pp. 1811-1817

Maser, C. & Trappe, J.M. (1984). *The seen and unseen world of the fallen tree. Gen. Tech. Re PNW-164. Portland, USA: Department of Agriculture, Forest Service*, Pacific Northwest Forest and Range Experiment Station

Maser, C.; Anderson, R.G. & Cromack, K. (1979). Dead and down woody material, In: *Wildlife Habitats in Managed Forests: The Blue Mountains of Oregon and Washington*, Thomas, J.W. (Ed.), USDA Forest Service, pp. 78-95

Mattson, K.G., Swank,W.T. & Waide, J.B. (1987). Decomposition of woody debris in a regenerating, clear-cut forest in the Southern Appalachians. *Canadian Journal of Forest Research*, Vol. 17, pp. 712-721

McMinn, J.W. & Crossley, D.A. (1993). *Biodiversity and coarse woody debris in southern forests.* USDA Forest Service. Report: SE-94

MCPFE (2002). Improved Pan-European indicators for sustainable forest management as adopted by the MCPFE Expert Level Meeting 2002. 27.04.2011, Available from http://www.mcpfe.org/system/files/u1/Vienna_Improved_Indicators.pdf

MCPFE (2007). *State of Europe´s forests 2007. The MCPFE report on sustainable forest management in Europe*. Liaison Unit Warsaw. 247 p., 20.01.2011, Available from http://www.mcpfe.org/filestore/mcpfe/Publications/pdf/state_of_europes_fores ts_2007.pdf

Merganič, J.; Šebeň, V. & Merganičová, K. (2011). Zásoba odumretého dreva v 1. až 8. lesnom vegetačnom stupni a na azonálnych stanovištiach, In: *Research of the classification methods and structural models of forest ecosystems favourable state – Assessment of the state and development with help of RS*, J. Vladovič (Ed.), NLC Zvolen, in press

Merganič, J.; Vorčák, J.; Merganičová, K.; Ďurský, J.; Miková, A.; Škvarenina, J.; Tuček, J. & Minďáš J. (2003). Diversity monitoring in mountain forests of Eastern Orava. EFRA, Tvrdošín, 200 p., 09.08.2011, Available from http://www.forim.sk/index_soubory/Merganic_Vorcak_Merganicova_Dursky_M ikova_Skvarenina_Tucek_Mindas_2003.pdf

Merganičová, K. & Merganič, J. (2010). Coarse woody debris carbon stocks in natural spruce forests of Babia hora. *Journal of Forest Science*, Vol., 56, No. 9, p. 397-405, ISSN: 1212-4834

Merganičová, K. (2004). *Modelling forest dynamics in virgin and managed forest stands.* Dissertation thesis, BOKU Vienna, 155 p.

Merganičová, K.; Merganič, J. & Vorčák, J. (2004). Zásoba odumretého dreva v NPR Babia hora. *Beskydy*, Vol. 17, pp. 137-142

Möller, G., (1994). Alt- und Totholzlebensräume. Ökologie, Gefährdungssituation, Schutzmaßnahmen. *Beiträge Forstwirtschaft und Landschaftsökologie*, Vol. 28, No. 1, pp. 7-15

Morelli, S.; Paletto, A. & Tosi, V. (2007). Deadwood in forest stands: assessment of wood basic density in some tree species, Trentino, *Italy.Forest*, Vol. 4, No. 4, pp. 395-406

Mössmer, R., (1999). Totholz messen im Staatswald. *LWF aktuell*, Vol. 18, pp. 7

Müller, J. & Bussler, H. (2008). Key factors and critical thresholds at stand scale for saproxylic beetles in a beech dominated forest, southern Germany. *Rev. Écol. (Terre Vie)*, Vol. 63, pp. 73-82

Müller, J. & Bütler, R. (2010). A review of habitat thresholds for dead wood: a baseline for management recommendations in European Forests European. *Journal of Forest Research*, Vol. 129, No. 6, pp. 981-992

Müller, J. & Schnell, A. (2003). Was lernen wir, wenn wir nichts tun? *LWF aktuell*, Vol. 40, pp. 8-11

Nadkarni, N.M. & Matelson, T.J. (1992). Biomass and nutrient dynamics of fine litter of terrestrially rooted material in a Neotropical montane forest, Costa Rica. *Biotropica*, Vol. 24, pp. 113-120

Næsset, E., (1999). Relationship between relative wood density of Picea abies logs and simple classification systems of decayed coarse woody debris. *Scandinavian Journal of Forest Research*, Vol. 14, pp. 454-461

Nakagawa, M.; Kurahashi, A. & Kaji, M. (2001). The effects of selection cutting on regeneration of Picea jezoensis and Abies sachalinensis in the sub-boreal forests of Hokkaido, northern Japan. *Forest Ecology and Management*, Vol. 146, pp. 15-23

Narukawa, Y. & Yamamoto, S. (2002). Effects of dwarf bamboo (Sasa sp.) and forest floor microsites on conifer seedling recruitment in a subalpine forest, Japan. *Forest Ecology and Management*, 163, pp. 61-70

Narukawa, Y.; Iida, S. & Tanouchi, H. (2003). State of fallen logs and the occurrence of conifer seedlings and saplings in boreal and subalpine old-growth forests in Japan. *Ecological Research*, 18, pp. 267-277

Nilsson, S.G.M.; Niklasson, J.; Hedin, G.; Aronsson, J.M.; Gutowski, P.; Linder, H.; Ljungberg, G. & Mikusinski-Ranius, T. (2002). Densities of large living and dead trees in old-growth temperate and boreal forests. *Forest Ecology and Management*, Vol. 161, pp. 189-204

Nordén, B.; Ryberg, M.; Gotmark, F. & Olausson, B. (2004). Relative importance of coarse and fine woody debris for the diversity of wood-inhabiting fungi in temperate broadleaf forests. *Biological Conservation*, Vol. 117, pp.1 – 10

Ódor, P. & Standovár, T. (2001). Richness of bryophyte vegetation in a near-natural and managed beech stands: the effects of management-induced differences in deadwood. *Ecological Bulletin*, Vol. 49, pp. 219-229

Oehmichen, K. (2007). *Erfassunf der Totholzmasse - Zusammenstellung von Verfahrensansätzen und Bewertung ihrer Eignung fuer massenstatistische Erhebungen*. Arbeitsbericht. Bundesforschungsanstalt fuer Forst- und Holzwirtschaft Hamburg, 24.08.2011, Available from http://www.bfafh.de/bibl/pdf/vii_07_1.pdf

Pasierbek, T.; Holeksa, J.; Wilczek, Z. & Żywiec, M. (2007). Why the amount of dead wood in Polish forest reserves is so small? *Nature Conservation*, Vol. 64, pp. 65-71

Pasinelli, K. & Suter, W. (2000). *Lebensraum Totholz*. Merkblatt fűr die Praxis. WSL Birmensdorf, Vol. 33, 24.08.2011, Available from http://www.wsl.ch/dienstleistungen/publikationen/pdf/5029.pdf

Persiani, A.M.; Audisio, P.; Lunghini, D.; Maggi, O.; Granito, V.M.; Biscaccianti, A.B.; Chiavetta, U. & Marchetti, M. (2010). Linking taxonomical and functional biodiversity of saproxylic fungi and beetles in broad-leaved forests in southern Italy with varying management histories. *Plant Biosystems*, Vol. 144, No. 1, pp. 250-261

Pesonen, A.; Maltamo, M.; Eerikäinen, K. & Packalén, P. (2008). Airborne laser scanning-based prediction of coarse woody debris volumes in a conservation area. *Forest Ecology and Management*, Vol. 255, pp. 3288-3296

Peterken, G.F. (1996). *Natural woodland: Ecology and conservation in northern temperate regions*. Cambridge, Cambridge University Press

Petráš, R. (1986). Mathematical model of stem shape. *Lesnícky časopis*, Vol. 32, No. 3, pp. 223-236

Petráš, R. (1989): Mathematical model of stem shape of coniferous tree species. *Lesnictví*, Vol. 35, No. 10, pp. 867-878

Petráš, R. (1990). Mathematical model of stem shape of broadleaved tree species. *Lesnícky časopis*, Vol. 36, No. 3, pp. 231-241

Pichler, V.; Homolák, M.; Skierucha, W.; Pichlerová, M.; Ramírez, D.; Gregor, J. & Jaloviar, P. (2011). Variability of moisture in coarse woody debris from several ecologically important tree species of the Temperate Zone of Europe. *Ecohydrology*. doi: 10.1002/eco.235

Pouska, V.; Svoboda, M. & Lepšová, A. (2010).The diversity of wood-decaying fungi in relation to changing site conditions in an old-growth mountain spruce forest, Central Europe. *European Journal of Forest Researsch*, Vol. 129, pp. 219–231

Pyle, C. & Brown, M.M. (1999). Heterogeneity of wood decay classes within hardwood logs. *Forest Ecology and Management*, Vol. 114, No. 2-3, pp. 253-259

Radtke, P.J.; Prisley, S.P.; Amateis, R.L.; Copenheaver, C.A. & Burkhart, H.E. (2004). A proposed model for deadwood C production and decay in loblolly pine plantations. (Special issue: Natural resource management to offset greenhouse gas emissions). *Environmental Management*, Vol. 33, Supplement 1, pp. 856-864

Radu, S. (2007). The ecological role of deadwood in natural forests, In: *Nature Conservation: Concept and Practice*, D. Gafta, & J. Akeroyd, (Eds.), Springer, Berlin, pp. 137–141

Rahman, M.M.; Frank, G.; Ruprecht, H. & Vacik, H. (2008). Structure of coarse woody debris in Lange-Leitn Natural Forest Reserve, Austria. *Journal of Forest Science*, Vol. 54, No. 4, pp. 161-169

Ranius, T.; Kindvall, O.; Kruys, N. & Jonsson, B.G. (2003). Modelling dead wood in Norway spruce stands subject to different management regimes. *Forest Ecology and Management*, Vol. 182, pp. 13–29

Rondeux, J. & Sanchez, C. (2009). *Review of indicators and field methods for monitoring biodiversity within national forest inventories*. Core variable: Deadwood. Environmental Monitoring and Assessment. Vol. 164, No. 1-4, pp. 617-630

Rouvinen, S.; Rautiainen, A. & Kouki, J. (2005). A Relation Between Historical Forest Use and Current Dead Woody Material in a Boreal Protected Old-Growth Forest in Finland. *Silva Fennica*, Vol. 39, No. 1, pp. 21-36

Sandoz, J.L. (1989). Grading of timber by ultrasound. *Wood Science Technology*, Vol. 23, pp. 95–108

Saniga, M. & Schütz, J.P. (2001). Dynamic changes in dead wood share in selected beech virgin forests in Slovakia within their development cycle. *Journal of Forest Science*, Vol. 47, No. 12, pp. 557-565

Scherzinger, W. (1996). *Naturschutz im Wald*. Qualitätsziele einer dynamischen Waldentwicklung. Praktischer Naturschutz. Verlag Eugen Ulmer, Stuttgart.

Schiegg, K. (2000). Effects of dead wood volume and connectivity on saproxylic insect species diversity. *Ecoscience*, Vol. 7, No. 3, pp. 290-298

Schmitt, M. (1992). Buchen-Totholz als Lebensraum für Xylobionte Käfer-Untersuchungen im Naturwaldreservat "Waldhaus" und zwei Vergleichsflächen im Wirtschaftswald (Forstamt Ebrach, Steigerwald). *Waldhygiene*, Vol. 19, pp. 97-191

Schreuder, H.T.; Gregoire, T.G. & Wood, G.B. (1993). *Sampling methods for multiresource forest inventory*. John Wiley 8 Sons, Inc., 446 p.

Schuck, A.; Meyer, P.; Menke, N.; Lier, M. & Lindner, M. (2004). Forest biodiversity indicator: dead wood - a proposed approach towards operationalising the MCPFE indicator. *EFI-Proceedings*, Vol. 51, pp. 49-77

Šebeň, V.; Kulla, L. & Jankovič, J. (2009). Analýza výskytu, množstva a štruktúry odumretého dreva na tatranskom kalamitisku, In: *Vplyv vetrovej kalamity na vývoj lesných porastov vo Vysokých Tatrách*, L. Tužinský, & J. Gregor, (Ed.), Zborník recenzovaných vedeckých prác, TU Zvolen, pp. 75-84

Shiver, B.D. & Borders, B.E. (1996). *Sampling techniques for forest resource inventory*. John Wiley 8 Sons, Inc., 356 p.

Siitonen, J. (2001). Forest management, coarse woody debris and saproxylic organisms: Fennoscandian boreal forests as an example. *Ecological Bulletin*, Vol. 49, pp. 11 – 42

Siitonen, J.; Martikainen, P.; Punttila, P.& Rauh, J. (2000). Coarse wood debris and stand characteristics in mature managed and old growth boreal mesic forests in southern Finland. *Forest Ecology and Management*, Vol. 128, pp. 211-225

Simila, M.; Kouki, J. & Martikainen, P. (2003). Saproxylic beetles in managed and seminatural scots pine forests: quality of dead wood matters. *Forest Ecology and Management*, Vol. 174, pp. 365-381

Šmelko, Š. & Merganič, J. (2008). Some methodological aspects of the National Forest Inventory and Monitoring in Slovakia. *Journal of Forest Science*, Vol. 54, No. 10, pp. 476-483, ISSN: 1212-4834

Šmelko, Š. (1968). *Matematicko – štatistická inventarizácia zásob lesných porastov.* SAV, Bratislava, 299 p.

Šmelko, S. (1985). *Nové smery v metodike a technike inventarizácie lesa.* Vedecké a pedagogické aktuality, 6, Vysoká škola lesnícka a drevárska Zvolen, 122 p.

Šmelko, Š. (2010). New methodical procedures for the quantification of deadwood and its components in forest ecosystems. *Lesnícky Časopis – Forestry Journal*, Vol. 56, No. 2, pp. 155-175

Šmelko, Š.; Merganič, J.; Šebeň, V.; Raši, R. & Jankovič, J. (2006). *Národná inventarizácia a monitoring lesov Slovenskej republiky 2005-2006.* Metodika terénneho zberu údajov (Pracovné postupy - 3. Doplnená verzia). Národné lesnícke centrum Zvolen, 129 p.

Smykała, J. (1992). Health and sanitary condition of the forests belonging to the State Forests in 1991. *Sylwan*, Vol. 136, No. 7, pp. 5-15

Sollins, P. (1982). Input and decay of coarse woody debris in coniferous forest stands in western Oregon and Washington. *Canadian Journal of Forest Research*, Vol. 12, pp. 18-28

Speight, M.C.D. (1989). *Saproxylic invertebrates and their conservation.* Nature and Environment Series, No. 42. Council of Europe, Strasbourg, France

Stachura, K.; Bobiec, A.; Obidziński, A.; Oklejewicz, K. & Wolkowycki, D. (2007). *Old trees and decaying wood In forest ecosystems of Poland "Old Wood".* A toolkit for participants, Version 07, 05.08.2011, Available from http://oldwood.dle.interia.pl/OW_07.pdf

Stevens, V. (1997). *The ecological role of coarse woody debris: an overview of the ecological importance of CWD in B.C. forests.* Res. Br., B.C. Min. For., Victoria, B.C. Work. 30: 26

Stokland, J.N.; Tomter, S.M. & Söderberg, U. (2004). Development of Dead Wood Indicators for Biodiversity Monitoring: Experiences from Scandinavia. *EFI Proceedings*, pp. 207-228

Storaunet, K.O. & Rolstand, J. (2002). Time since death and fall of Norway spruce logs in old-growth and selectively cut boreal forest. *Canadian Journal of Forest Research*, Vol. 32, pp. l80l - 1812

Szewczyk, J. & Szwagrzyk, J. (1991). Tree regeneration on rotten wood and on soil in old-growth stand. *Vegetatio*, Vol. 122, No. 1, pp. 37-46

Takahashi, M.; Sakai, Y.; Ootomo, R. & Shiozaki, M. (2000). Establishment of tree seedlings and water-soluble nutrients in coarse woody debris in an old-growth Picea-Abies forest in Hokkaido, northern Japan. *Canadian Journal of Forest Research*, Vol. 30, pp. 1148-1155

Thomas, J.W. (2002). *Dead Wood: From Forester's Bane to Environmental Boom.* USDA Forest Service Gen. Tech. Rep. PSW-GTR-181

Thompson, S.K. (2002). *Sampling*. John Wiley 8 Sons, Inc., 367 p.

Travaglini, D. & Chirici, G. (2006). *ForestBIOTA project. Forest Biodiversity Test-phase Assessments: Deadwood assessment*. Work report. Accademia Italiana di Scienze Forestali 19 p., 10.01.2011, Accessed from http://www.forestbiota.org/docs/report_DEADWOOD.pdf

Travaglini, D.; Barbati, A.; Chirici, G.; Lombardi, F.; Marchetti, M. & Corona, P. (2007). ForestBIOTA data on deadwood monitoring in Europe. *Plant Biosystems*, Vol. 141, No. 2, pp. 222-230

Turner, D.P.; Koerper, G.J.; Harmon, M.E. (1995). A carbon budget for forests of the conterminous United States. *Ecological Applications*, Vol. 5, pp. 421-436

Ulbrichová, I.; Remeš, J. & Zahradník, D. (2006). Development of the spruce natural regeneration on mountain sites in the Šumava Mts. *Journal of Forest Science*, Vol. 52, pp. 446-456

Vacik, H.; Rahman, M.M.; Ruprecht, H. & Frank, G. (2009). Dynamics and structural changes of an oak dominated Natural Forest Reserve in Austria. *Botanica Helvetica*, Vol. 119, pp. 23-29

Vallauri, D.; Andre, J. & Blondel, J. (2003). Dead wood - a typical shortcoming of managed forests. *Revue Forestiere Francaise*, Vol. 55, No. 2, pp. 99-112

Vandekerkhove, K.; Keersmaeker, De L.; Menke, N.; Meyer, P. & Verschelde, P. (2009). When nature takes over from man: Dead wood accumulation in previously managed oak and beech woodlands in North-western and Central Europe. *Forest Ecology and Management*, Vol. 258, pp. 425-435

Veerkamp, M.T.; 2003). The importance of large dead beech wood for fungi. Nederlands-Bosbouwtijdschrift. Vol. 75, No. 5, pp. 10-14

Vehmas, M; Packalén, P. & Maltamo, M. (2009). *Assessing deadwood existence in canopy gaps by using ALS data*. Silvilaser 2009, October 14-16, 2009 – College Station, Texas, USA

Vorčák, J.; Merganič, J. & Merganičová, K. (2005). Deadwood and spruce regeneration. *Lesnická práce*, Vol. 5, pp. 18-19

Vorčák, J.; Merganič, J. & Saniga, M. (2006). Structural diversity change and regeneration processes of the Norway spruce natural forest in Babia hora NNR in relation to altitude. *Journal of Forest Science*, Vol. 52, No. 9, pp. 399-409

Vrška, T.; Hort, L.; Odehnalová, P.; Adam, D. & Horal, D. (2001a). The Milešice virgin forest after 24 years (1972-1995). *Journal of Forest Science*, Vol. 47, No. 6, pp. 255-276

Vrška, T.; Hort, L.; Odehnalová, P.; Adam, D. & Horal, D. (2001b). The Razula virgin forest after 23 years (1972-1995). *Journal of Forest Science*, Vol. 47, No. 1, pp. 15-37

Vrška, T.; Hort, L.; Odehnalová, P.; Adam, D. & Horal, D. (2001c) The Boubín virgin forest after 24 years (1972-1996) – development of tree layer. *Journal of Forest Science*, Vol. 47, No. 10, pp. 439-459

Wagner van, C.E. (1968). The line intersect method in forest fuel sampling. *Forest Science*, Vol. 14, pp. 20-26

Warren, W.G. & Olsen, P.F. (1964). A line intersect technique for assessing logging waste. Forest Science, Vol. 13, pp. 267-276

Weiss, P.; Schieler, K.; Schadauer, K.; Radunsky, K. & Englisch, M. (2000). *Carbon budget of Austrian forests and considerations about Kyoto Protocol*. Monographie 106. Federal Environment Agency, Wien (in German).

Winter, S.; Fischer, H.S. & Fischer, A. (2010). Relative quantitative reference approach for naturalness assessments of forests. *Forest Ecology and Management*, Vol. 259, pp. 1624-1632

Woodall, C. & Williams, M.S. (2005). Sampling protocol, estimation, and analysis procedures for the down woody materials indicator of the FIA program. *General-Technical-Report-North-Central-Research-Station,-USDA-Forest-Service.* (NC-256), 47 p.

Woodall, C.W. & Liknes, G.C. (2008). Climatic regions as an indicator of forest coarse and fine woody debris carbon stocks in the United States. *Carbon Balance and Management*, Vol. 3, pp. 5

Yatskov, M. (2001). *Chonosequence of wood decomposition in the boreal forests of Russia.* [PhD. Thesis.], 20.01.2010, Available from
http://andrewsforest.oregonstate.edu/pubs/webdocs/reports/wood_decomp.ht

Yin, X. (1999). The decay of forest woody debris: numerical modeling and implications based on some 300 data cases from North America. *Oecologia*, Vol. 121, pp. 81-98

Zhou, L.; Dai, L.; Gu, H. & Zhong, L. (2007). Review on the decomposition and influence factors of coarse woody debris in forest ecosystem. *Journal of Forestry Research*, Vol. 18, pp. 48-54

Zielonka, T. & Piatek, G. (2004). The herb and dwarf shrubs colonization of decaying logs in subalpine forest in the Polish Tatra Mountains. *Plant Ecology*, Vol. 172, pp. 63-72

Zielonka, T. (2006). Quantity and decay stages of coarse woody debris in old-growth subalpine spruce stands of the western Carpathians, Poland. *Canadian Journal of Forest Research*, Vol. 36, pp. 2614-2622

Zimmerman, J.K.; Pulliam, W.M.; Lodge, D.J.; Quinones-Orfila, V.; Fetcher, N.; Guzman-Grajales, S.; Parrotta, J.A.; Asbury, C.E.; Walker, L.R. & Waide, R.B. (1995). Nitrogen immobilization by decomposing woody debris and the recovery of tropical wet forest from hurricane damage. *Oikos*, Vol. 72, No. 3, pp. 314-322

Zöhrer, F. (1980). *Forstinventur: Ein Leitfaden für Studium and Praxis.* Hamburg-Berlin

Plant Diversity of Forests

Ján Merganič[1,2], Katarína Merganičová[1,2],
Róbert Marušák[1] and Vendula Audolenská[1]
[1]Czech University of Life Sciences in Prague,
Faculty of Forestry, Wildlife and Wood Sciences,
Department of Forest Management, Praha
[2]Forest Research, Inventory and
Monitoring (FORIM), Železná Breznica
[1]Czech Republic
[2]Slovakia

1. Introduction

Changes in biological diversity of natural ecosystems have in the second half of 20th century become a global problem due to intensive human activities. Therefore, higher attention has been paid to these problems. The year 1992 can be considered as the pivotal year in this field since in this year the Convention on Biological Diversity was approved on the United Nations Conference on Environment and Development in Rio de Janeiro. This document defines biological diversity - biodiversity as „the variety and variability among living organisms from all sources including inter alia, terrestrial, marine and other aquatic ecosystems and the ecological complexes of which they are part". This definition covers three fundamental components of diversity: genetic, species, and ecosystem diversity (Duelli, 1997, as cited in Larsson, 2001; Merganič & Šmelko, 2004). However, also this widely accepted definition like many others fails to mention ecological processes, such as natural disturbances, and nutrient cycles, etc., that are crucial to maintaining biodiversity (Noss, 1990). The complexity of the understanding of the term biodiversity was well documented by Kaennel (1998). Therefore, Noss (1990) suggested that for the assessment of the overall status of biodiversity more useful than a definition would be its characterisation that identifies its major components at several levels of organisation. Franklin et al. (1981 as cited in Noss, 1990) recognised three primary attributes of ecosystems: composition, structure, and function.

1.1 Species diversity

Strictly speaking, species diversity is the number of different species in a particular area (species richness) weighted by some measure of abundance such as number of individuals or biomass. However, conservation biologists often use the term species diversity even when they are actually referring to species richness, i.e. to number of present species (Harrison et al., 2004). Noss (1990) defines species diversity as a composition that refers to the identity and variety of elements in a population, includes species lists and measures of species diversity and genetic diversity.

1.2 Structural diversity

"Structural diversity refers to the physical organisation or pattern of a system, including the spatial patchwork of different physical conditions in a landscape, habitat mosaics, species assemblages of different plant and animal communities, and genetic composition of subpopulations" (Stokland et al., 2003). The main structural indicators that are used to describe the conditions for forest biodiversity include stand vertical structure, age class distribution and the amount of dead wood (Christensen et al., 2004). They represent an indirect approach „as they show, typically on a rather gross scale, how the house is built, but give no information on whether the inhabitants have moved in" (Christensen et al., 2004).

1.3 Functional diversity

According to Noss (1990), function involves all ecological and evolutionary processes, including gene flow, disturbances, and nutrient cycling. "Functional diversity involves processes of temporal change, including disturbance events and subsequent succession, nutrient recycling, population dynamics within species, various forms of species interactions, and gene flow" (Stokland et al., 2003).

2. Factors influencing plant diversity in forest ecosystems

Within certain time and space, diversity is determined by the combination of abiotic constraints, biotic interactions, and disturbances (Frelich et al., 1998; Misir et al., 2007; Nagaraja et al., 2005; Spies & Turnier, 1999; Ucler et al., 2007). Abiotic factors, such as elevation, slope, aspect, soil texture, climate etc., specify the conditions of physical environment and thus the primary species distribution. The relations were already regarded and studied in 19th century (Hansen & Rotella, 1999). The parameters affecting the plant growth and nutrient availability, e.g. climate, are considered as primary factors (Terradas et al., 2004), while terrain characteristics, e.g. elevation, are regarded as indirect factors, because they do not influence the plant growth directly, but are correlated to primary factors (Pausas et al., 2003; Bhattarai et al., 2004).

Primary climate and site conditions have influenced and determined biodiversity on a specific site in the long-term development of forest ecosystems (Stolina, 1996). Hence, the actual biodiversity is the result of the adaptation process of species. In the current conditions of climate change the species will have to respond to faster changes. Although the effect of climate change will vary from site to site, it is likely that its impacts on ecosystems will be adverse, as species will have to deal with a variety of new competitors, and biotic factors (diseases, predators), to which they have no natural defense so far (IUCN, 2001).

Indirect factors are often used in the analyses, when the information about the primary factors is not available (Pausas & Saez, 2000). Most often, the relationship between the diversity and elevation is examined (Bachman et al., 2004; Bhattarai & Vetaas, 2003; Grytnes & Vetaas, 2002), while the influence of other topography characteristics is tested only seldom (Johnson, 1986; Palmer et al., 2000). Although modern ecologists focus mainly on other influencing factors, e.g. natural disturbances, the influence of abiotic conditions on species diversity has recently begun to gain attention of researchers (Austin et al., 1996; Burns, 1995; Hansen & Rotella, 1999; Ohmann & Spies, 1998; Rosenzweig, 1995). However, most of these works analyse the environmental factors only with regard to the number of tree species representing just one part of species diversity.

Abiotic factors, such as elevation, slope, aspect, terrain type etc., create together a unique complex of environmental conditions specifying forest communities (Spies & Turner, 1999). The relation between elevation and species diversity is generally accepted and was documented by several authors, not only for tree species but also regarding the diversity of plants and animals (Rosenzweig, 1995). Very often hump-shaped curves with maximum species diversity at mid-elevations were reported (Bhattarai & Vetaas, 2003; Bachman et al., 2004; Ozcelik et al., 2008). In Merganič et al. (2004), elevation was also found to have a significant influence on tree species diversity, but at mid-elevations the lowest values of tree species diversity were observed. This performance can most probably be explained by the fact that in the Slovak Republic at about 600 m above sea level, beech has its optimum growing conditions, which causes that at these altitudes beech is so vital and competitive that other species become rare. Johnson (1986) and Ozcelik et al. (2008) detected the significant correlation between tree species diversity and aspect.

2.1 Forest management

In Europe forests have played an important role since their establishment after the last ice age that ended 12,000 years ago. In the human thoughts, the forest was an unknown and untouched place with secrets and dangers. It provided a man with a shelter, fuel wood, and cosntruction material (Reinchholf, 1999). A man started to have a stronger influence on a forest ecosystem around the year 4,000 B.C. The impact was first low; he cut trees to obtain space for settlements and for grazing of his animals. With the increasing demands on space, forest ecosystems were more and more utilised, which led to the significant decrease of forest area in the whole Europe. In 16th century, the first attempts to grow introduced tree species, namely Castanea sativa, occurred. However, the most significant changes of forest ecosystems started in 19th century with the beginning of a so-called "spruce and pine mania". In this period, the majority of forestland was afforested with spruce, even in completely unsuitable conditions. The main reason of this boom was to maximise wood production. Nowadays, it is known that such an approach has had a negative impact on stand stability, as well as on forest biodiversity. The look of the forests today particularly in the densely inhabited areas is related to management intensity and methods (Hédl & Kopecký, 2006). The absence of suitable management is another cause of decreasing forest biodiversity (Hédl, 2006).

Although currently biodiversity has become a key component of Central European forests, there is only a limited number of studies, which examine the influence of forest management on biodiversity of e.g. plants (Prevosto, 2011). In addition, the results are often contradictory. On one side, some works present that forest management has a negative effect on biodiversity (Gilliam & Roberts, 1995; Sepp & Liira, 2009). Other works (e.g. Battles, 2001; Newmaster, 2007; Ramovs & Roberts, 2005; Ravindranath, 2006; Wang & Chen, 2010) show that a well-chosen management can influence biodiversity positively. The compatibility of suitable management activities with biodiversity conservation is critical to ensure wood harvesting and other ecologically valuable aspects in forested land (Eriksson & Hammer, 2006). Sustainable forest management represents how high biodiversity can be achieved together with high wood production. This type of management maintains forests and forest soil in order to secure biodiversity, productivity, regeneration capacity, vitality, and abilities to fulfil all ecological, economic, and social functions today and in future on any spatial scale (local, regional, national) without the drawback on other ecosystems (Poleno, 1997). Sustainability means the ability to provide current and future generations with permanent and optimal wood yield and other forest ecosystem products (Smola, 2008).

3. Diversity assessment

Due to the complexity of biodiversity and of forest ecosystems, complete assessments of biodiversity are not practically achievable (Humphrey & Watts, 2004) because of the impossibility to monitor all taxa or features (Lindenmayer, 1999). Therefore, means to reduce complexity are necessary (Christensen et al., 2004). In this context, reliable indicators or short-cut measures of biodiversity are searched for (Ferris & Humphrey, 1999; Jonsson & Jonsell, 1999; Noss, 1999; Simberloff, 1998 as cited in Humphrey & Watts, 2004). From the long-term perspective, the basic criterion for any biodiversity assessment system is that it is based on an enduring set of compositional, structural and functional characteristics (Allen et al., 2003). In addition, a complete long-term biodiversity strategy must take into account both interactions between the different geographical levels and the fact that different elements of biodiversity are dependent on different geographical scales, in different time perspectives (Larsson, 2001).

3.1 Species diversity

Species diversity can be evaluated by a great number of different methods (e.g. see Krebs, 1989; Ludwig & Reynolds, 1988). All of the proposed methods are usually based on at least one of the following three characteristics (Bruciamacchie, 1996):

- species richness – the oldest and the simplest understanding of species diversity expressed as a number of species in the community (Krebs, 1989);
- species evenness – a measure of the equality in species composition in a community;
- species heterogeneity – a characteristic encompassing both species abundance and evenness.

The most popular methods for measurement and quantification of species diversity are species diversity indices. During the historical development, the indices have been split into three categories: indices of species richness, species evenness and species diversity (Krebs, 1989; Ludwig & Reynolds, 1988). The indices of each group explain only one of the above-mentioned components of species diversity (Merganič & Šmelko, 2004).

3.1.1 Species richness

The term species richness was introduced by McIntosh (1967) to describe the number of species in the community (Krebs, 1989). Surely, the number of species S in the community is the basic measure of species richness, defined by Hill (1973) as diversity number of 0th order, i.e. N0. The basic measurement problem of N0 is that it is often not possible to enumerate all species in a population (Krebs, 1989). In addition, S depends on the sample size and the time spent searching, due to which its use as a comparative index is limited

(Yapp, 1979). Hence, a number of other indices independent of the sample size have been proposed to measure species richness. These indices are usually based on the relationship between S and the total number of individuals observed (Ludwig & Reynolds, 1988). Two such well-known indices are R1 and R2 proposed by Margalef (1958) and Menhinick (1964), respectively. Hubálek (2000), who examined the behaviour of 24 measures of species diversity in a data from bird censuses, assigned to the category of species richness-like indices also the index α (Fischer et al., 1943; Pielou, 1969), Q (Kempton & Taylor, 1976, 1978), and R500 (Sanders, 1968; Hurlbert, 1971).

3.1.2 Species evenness

Lloyd & Ghelardi (1964) were the first who came with idea to measure the evenness component of diversity separately (Krebs, 1989). The principle of the evenness measures is to quantify the unequal representation of species against a hypothetical community in which all species are equally common. Ludwig & Reynolds (1988) present five evenness indices E1 (Pielou, 1975, 1977), E2 (Sheldon, 1969), E3 (Heip, 1974), E4 (Hill, 1973), and E5 (Alatalo, 1981), each of which may be expressed as a ratio of Hill's numbers. The most common index E1, also known as J' suggested by Pielou (1975, 1977) expresses H' relative to maximum value of H' (= log S). Index E2 is an exponentiated form of E1. Based on the analysis of Hubálek (2000), McIntosh's diversity D (McIntosh, 1967; Pielou, 1969), McIntosh's evenness DE (Pielou, 1969), index J of Pielou (1969) and G of Molinari (1989), are also evenness measures.

3.1.3 Species heterogeneity

This concept of diversity was introduced by Simpson (1949) and combines species richness and evenness. The term heterogeneity was first applied to this concept by Good (1953). Many ecologists consider this concept to be synonymous with diversity (Hurlbert, 1971, as cited in Krebs, 1989). According to Peet (1974, as cited in Ludwig & Reynolds, 1988), an infinite number of diversity indices exist. Simpson proposed the first heterogeneity index λ, which gives the probability that two individuals picked at random from the community belong to the same species. This means that if the calculated probability is high, the diversity of the community is low (Ludwig & Reynolds, 1988). To convert this probability to a diversity measure, the complement of Simpson's original measure, i.e. $1-\lambda$, is used (Krebs, 1989).

Probably the most widely used heterogeneity index is the Shannon index H' (or Shannon-Wiener function), which is based on information theory (Shannon & Weaver, 1949). It is a measure of the average degree of "uncertainty" in predicting to what species an individual chosen at random from a community will belong (Ludwig & Reynolds, 1988). Hence, if H' = 0, the community consists of only one species, whereas H' is maximum (= log(S)) if all species present in the community are represented by the same number of individuals. Shannon index places most weight on the rare species in the sample, while Simpson index on the common species (Krebs, 1989).

From other heterogeneity measures we mention Brillouin Index H (Brillouin, 1956), which was first proposed by Margalef (1958) as a measure of diversity. This index is preferred being applied to data in a finite collection rather than H'. However, if the number of individuals is large, H and H' are nearly identical (Krebs, 1989). The indices N1 and N2 from Hill's family of diversity numbers (Hill, 1973), which characterise the number of "abundant", and "very abundant" species, respectively, also belong to diversity measures. The McIntosh index is based on the representation of a sample in an S-dimensional hyperspace, where each dimension refers to the abundancy of a particular species (Bruciamacchie, 1996). According to the evaluation performed by Hubálek (2000), NMS "number of moves per specimen" proposed by (Fager, 1972), H´adj, which is an adjusted H´ by the d(H) correction (Hutcheson, 1970), and R100 (Sanders, 1968; Hurlbert, 1971) can also be regarded as heterogeneity indices.

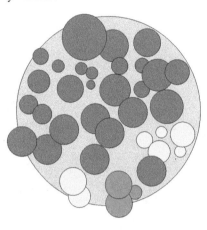

Legend:
◉ Fagus sylvatica, ○ Picea abies, ◎ Abies alba
Species richness: 3, Species evenness: Low, Species heterogeneity: Low.

Fig. 1. Assessment of tree species diversity.

3.2 Structural diversity

Structural diversity is defined as the composition of biotic and abiotic components in forest ecosystems (Lexer et al., 2000), specific arrangement of the components in the system (Gadow, 1999) or as their positioning and mixture (Heupler, 1982 as cited in Lübbers, 1999). According to Zenner (1999) the structure can be characterised horizontally, i.e. the spatial distribution of the individuals, and vertically in their height differentiation. Gadow & Hui (1999) define the structure as spatial distribution, mixture and differentiation of the trees in a forest ecosystem.

There exist a number of different methods to describe the structure and its components. The classical stand description is based on qualitative description of stand closure, mixture, density, etc. Graphical methods presenting diameter distribution, stand height distribution curves, tree maps, etc. are also useful. However, both verbal and graphical methods may not be sufficient to reveal subtle differences (Kint et al., 2000). Therefore, a number of quantitative methods have been proposed that should overcome these shortages. Partial reviews can be found in Pielou (1977), Gleichmar & Gerold (1998), Kint et al. (2000), Füldner (1995), Lübbers (1999), Gadow & Hui (1999), Neumann & Starlinger (2001), Pommerening (2002) etc.

3.2.1 Horizontal diversity

The indices characterising forest horizontal structure usually compare a hypothetical spatial distribution with the real situation (Neumann & Starlinger, 2001). Probably the most well-known index is the aggregation index R proposed by Clark & Evans (1954) that describes the horizontal tree distribution pattern (or spacing as named by Clark & Evans (1954), or positioning as defined by Gadow & Hui (1999)). It is a measure of the degree to which a forest stand deviates from the Poisson forest, where all individuals are distributed randomly (Tomppo, 1986). It is the ratio of the observed mean distance to the expected mean distance if individuals were randomly distributed.

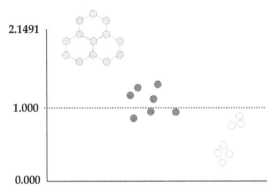

Fig. 2. Schematic visualisation of the assessment of forest horizontal structure using R index by Clark & Evans (1954).

A similar measure is the Pielou index of nonrandomness (Pielou, 1959), which quantifies the spatial distribution of trees by the average minimum distance from random points to the nearest tree (Neumann & Starlinger, 2001). The Cox index of clumping (Strand, 1953; Cox, 1971) is the ratio of variance to mean stem number on sub-plots. Gadow et al. (1998) proposed an index of neighbourhood pattern based on the heading angle to four next trees. Another commonly used measures of horizontal structure are methods proposed by Hopkins (1954), Prodan (1961), Köhler (1951) and Kotar (1993 as cited in Lübbers, 1999).

According to Gadow & Hui (1999), mixture is another component of structure. For the quantification of mixing of two tree species, Pielou (1977) proposed the segregation index based on the nearest neighbour method like the index A of Clark & Evans, while the calculated ratio is between the observed and expected number of mixed pairs under random conditions. Another commonly used index is the index DM (from German Durchmischung) of Gadow (1993) adjusted by Füldner (1995). On the contrary to the segregation index, DM accounts for multiple neighbours (Gadow, 1993 used 3 neighbours) and is not restricted to the mixture of two species (Kint et al., 2000).

Differentation is the third component of structure (Gadow & Hui, 1999), which describes the relative changes of dimensions between the neighbouring individuals (Kint et al., 2000). Gadow (1993) and Füldner (1995) proposed the differentiation index T, which is an average of the ratios of the smallest over the largest circumference calculated for each tree and its n nearest neighbours. Instead of the circumference, diameter at breast height can be used in this index to describe the horizontal differentiation as presented by Pommerening (2002). Values of the index T close to 0 indicate stands with low differentiation, since neighbouring trees are of similar size. Aguirre et al. (1998) and Pommerening (2002) suggested the scales of five or four categories of differentiation, respectively.

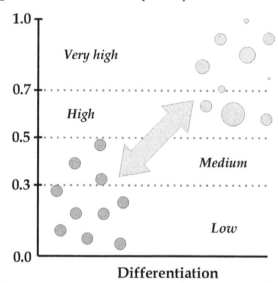

Fig. 3. Schematic visualisation of the assessment of stand differentiation according to Füldner (1995) with categories proposed by Aguirre et al. (1998).

3.2.2 Vertical diversity

While there are many indices that measure horizontal structure, there are only few for vertical structure (Neumann & Starlinger, 2001). Simple measures such as the number of vegetation layers within a plot can be used as an index of vertical differentiation (Ferris-Kaan & Patterson, 1992 as cited in Kint et al., 2000). The index A developed by Pretzsch (1996, 1998) for the vertical species profile is based on the Shannon index H′. In comparison with H′ the index A considers species portions separatelly for a predefined number of height layers (Pretzsch distinguished 3 layers). The index proposed by Ferris-Kaan et al. (1998) takes the cover per layer into account, but needs special field assessments (Neumann & Starlinger, 2001). Therefore, using the same principles as Pretzsch (1996), i.e. Shannon index and stratification into height layers, Neumann & Starlinger (2001) suggested an index of vertical evenness VE that characterises the vertical distribution of coverage within a stand. The differentiation index T of Gadow (1993) is also applicable for the description of vertical differentiation, if the index is calculated from tree heights.

3.2.3 Complex diversity

Complex indices combine several biodiversity components in one measure. These indices are usually based on an aditive approach, i.e. the final value is obtained as a sum of the values of individual biodiversity components. Usually, two ways of quantification individual biodiversity components are applied: (1) by assigning the value on the base of a pre-defined scale, or (2) to use the real measurement units. In addition, if required individual biodiversity components can be assigned different weights according to their importance for the whole biodiversity.

single-storey   double-storey   multiple-storey

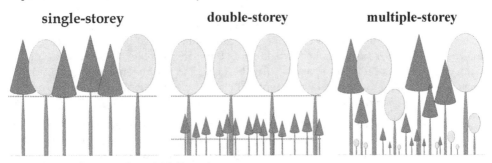

Fig. 4. Quantification of vertical diversity.

The first system of biodiversity assessment using scores is simple and easy to use (Meersschaut & Vandekerkhove, 1998). Such quantification was already used in 1969 by Randwell to assess the need for the protection of seashore sites on the base of Comparative Biological Value Index (Nunes et al., 2000). Meersschaut & Vandekerkhove (1998) developed a stand-scale forest biodiversity index based on available data from forest inventory. The index combines four major aspects of a forest ecosystem biodiversity: forest structure, woody and herbaceous layer composition, and deadwood. Each aspect consists of a set of indicators, e.g. forest structure is defined by canopy closure, stand age, number of stories, and spatial tree species mixture. The indicators are given a score determined on the basis of a common agreement. The biodiversity index is calculated as the sum of all scores, while its maximum value is set to 100. Another complex index named Habitat Index HI was developed by Rautjärvi et al. (2005). The authors also use the name habitat index model as it was produced as a spatial oriented model. The inputs in the model come from thematic maps from Finnish Multi-source national forest inventory (predicted volume of growing stock, predicted stand age, and predicted potential productivity) and kriging interpolation maps from national forest inventory plot data (volume of dead wood, and a measure for naturalness of a stand). The input variables were selected based on the forest biodiversity studies in Scandinavia. The index is of additive form where all input layers contribute to the result layer. All input variables (layers) are reclassified and enter the model as discreet variables, while each input layer is assigned a different weight according to its importance to biodiversity.

The second quantification method was used in the model BIODIVERSS proposed by Merganič & Šmelko (2004) that estimates tree species diversity degree of a forest stand by summing up the values of 5 diversity indices (R1, R2, λ, H' and E1). The fundamental method of the model BIODIVERSS is a predictive discriminant analysis (StatSoft Inc., 2004; Huberty, 1994; Cooley & Lohnes, 1971), which means that each species diversity degree is represented by one discriminant equation. For each examined forest stand, four discriminant scores are calculated, and the stand is assigned a species diversity degree with maximum discriminant score.

LLNS index proposed by Lähde et al. (1999) is a complex index for calculating within-stand diversity using the following indicator variables: stem distribution of live trees by tree species, basal area of growing stock, volume of standing and fallen dead trees by tree species, occurrence of special trees (number and significance), relative density of undergrowth, and volume of charred wood. The LLNS index is calculated as the sum of diversity indices describing particular components (i.e. living trees, dead standing trees etc.). However, the authors also developed a scoring table for the indicator variables. The final value of LLNS is then obtained by adding all the scores together. The evaluation of this index using Finnish NFI data revealed, that the LLNS index differentiates even-sized and uneven-sized stand structures, the development classes of forest stands and site-types fairly well (Lähde et al., 1999).

A special category of complex indices covers complex structural indices that encompass several components of structural diversity. For example, Jaehne & Dohrenbusch (1997) proposed the Stand Diversity Index that combines the variation of species composition, vertical structure, spatial distribution of individuals and crown differentiations. The Complexity Index by Holdridge (1967) is calculated by multiplying four traditional measures of stand description: dominant height, basal area, number of trees and number of species. Hence, this index contains no information on spatial distribution nor accounts for within stand variation (Neumann & Starlinger, 2000). Zenner (1999), and Zenner & Hibbs (2000) developed the Structural Complexity Index (SCI) based on the vertical gradient differences between the tree attributes and the distances between the neighbouring trees. When all trees in a stand have the same height, the value for SCI is equal to one, which is the lower limit of SCI (Zenner & Hibbs, 2000).

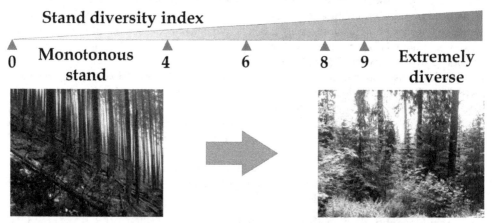

Fig. 5. Complex assessment of stand diversity according to Jaehne & Dohrenbusch (1997).

3.3 Functional diversity

From a functional point of view, species can be subdivided into categories like primary producers, herbivores, predators, and decomposers (Stokland et al., 2003). Belaoussoff et al. (2003) defined a functional group as a group of species, which do not necessarily have to be related, but which exploit a common resource base in a similar fashion. Hence, there is an overlap in resource requirements between species in a functional group (Belaoussoff et al., 2003).

The BEAR-project strongly recommends including functional indicators in any Biodiversity Evaluation Tool. Within the framework of the BEAR project, fire, wind and snow, and biological disturbance have been identified as the most important functional key factors in the group of "natural influences", while the area affected by a particular factor are suggested as possible indicators with high ecological significance (BEAR Newsletter 3). Although in reality the ecosystem function might be more important than species diversity (Sobek & Zak, 2003), structural and compositional indicators are considered to be more tractable for end-users (Angelstam et al., 2001 as cited in Humphrey & Watts, 2004).

3.4 Diversity inventory in forest ecosystems

To get an overview, forest inventory data can be a cost effective source of information for large areas (Söderberg & Fridman, 1998), because forest inventories represent a major source of data concerning forests (Estreguil et al., 2004). The original aim of forest inventories was to describe the main features of forests in terms of size, condition, and change, particularly from the production perspective (Rego et al., 2004). An increasing demand for information on non-productive functions of forests caused that recently variables more related to biodiversity have been introduced to forest inventories (Söderberg & Fridman, 1998). For example, the recognition of the ecological importance of decaying wood has led to the incorporation of quantitative measures of deadwood in forest inventories (Humphrey et al., 2004). Hence, national forest inventories are becoming more comprehensive natural resources surveys (Corona & Marchetti, 2007) .

Basically, forest inventories provide us with the information about: 1) forest area and land cover, 2) resource management (growing stock), 3) forestry methods and land use (felling systems, regeneration methods, road network density), 4) forest dynamics with regard to different disturbance factors (fire, storm, insect, browsing), 5) forest state (tree species composition, age distribution, dimension of living trees, tree mortality and deadwood), and partly also about 6) conservation measures, i.e. protected forest areas (Stokland et al., 2003). Hence, data from forest inventories are also useful for biodiversity assessment. For example, these data can be used for the quantification of several biodiversity indicators related to species composition, mainly in terms of species richness and the presence of species of high conservation value (threatened or endemic species, Corona & Marchetti, 2007).

However, data from forest inventories may not be suitable for every analysis. For example, national forest inventory field plots are inadequate for measuring landscape patterns of structural ecosystem diversity because of the small plot size (Stokland et al. 2003). In addition, in many cases precision guidelines for the estimates of many variables cannot be satisfied due to budgetary constraints and natural variability among plots (McRoberts et al. 2005). In neither of the cases, it is efficient to increase the plot size or their number. Instead, other data sources that enable rapid data generation, e.g. digital photogrammetry; geographical information systems (GIS), digital elevation model (DEM), global positioning system (GPS) or remote sensing (Gallaun et al., 2004; Kias et al., 2004 as cited in Wezyk et al., 2005) can be used more efficiently. Fieldwork itself has been enhanced by satellite positioning systems (GPS), automatic measuring devices, field computers and wireless data transfer (Holopainen et al., 2005).

For special purposes, specific monitoring programmes are needed. These programmes attempt to investigate particular features of a forest ecosystem that are of specific interest and their monitoring is not included within national forest inventories. Many of such surveys have been performed by non-governmental organisations and within the frame of specific forest monitoring programmes (Heer et al., 2004). Although this kind of information can be of high value at a local or national scale, its applicability at a higher level (region, Europe) is restricted and requires pre-processing of data with regard to their quality, and biases and gaps in time and space (Heer et al., 2004). Therefore, many international projects dealing with biodiversity have been solved in the last decade (e.g. BioAssess, BEAR, ForestBIOTA, ALTER-Net, SEBI, DIVERSITAS).

The quantification of biodiversity indicators can be performed in two ways, which affect the calculation of their confidence intervals. One method is that the indicator is calculated from the summary data about the whole population. In this case, there are several possibilities how to obtain the summary data:

- by accurate measurement of all individuals in population, i.e. complete survey;
- by visual estimation during the inspection of the examined population;
- by sampling methods in such a way, that the summary information is obtained by summing up the data collected on several places in a forest stand.

We call this approach the "method of sum". Biodiversity indicator determined with this method refers to the area that is larger than the minimum area. Hence, the comparison of the results of different populations is usually correct. In other cases, it is possible to use various standardisation methods given in e.g. Ludwig & Reynolds (1988) or Krebs (1989).

The second approach is called the "method of mean", because in this case biodiversity indicators are determined on several locations distributed over the whole community, and from them the average value typical for the whole population is derived. An important condition of this method is to assess biodiversity indicators on the samples of equal size, because in this case area has a significant effect on the value of biodiversity indicator. The final value of the biodiversity indicator refers to the area of the samples. Another alternative of this method is to determine indicator on the same number of individuals, e.g. 20 trees (Merganič & Šmelko, 2004).

At the ecosystem and landscape level, remote sensing represents a powerful and useful tool for biodiversity assessment (Ghayyas-Ahmad, 2001; Innes & Koch, 1998; Foody & Cutler, 2003). This method can provide cost efficient spatial digital data which is both spatially and spectrally more accurate than before (Holopainen et al., 2005). Moreover, remote sensing technology can provide the kind of information that was previously not available to forestry at all or was not available on an appropriate scale (Schardt et al., 2005). According to Innes & Koch (1998), "remote sensing provides the most efficient tool available for determining landscape-scale elements of forest biodiversity, such as the relative proportion of matrix and patches and their physical arrangement. At intermediate scales, remote sensing provides an ideal tool for evaluating the presence of corridors and the nature of edges. At the stand scale, remote sensing technologies are likely to deliver an increasing amount of information about the structural attributes of forest stands, such as the nature of the canopy surface, the presence of layering within the canopy and presence of coarse woody debris on the forest floor."

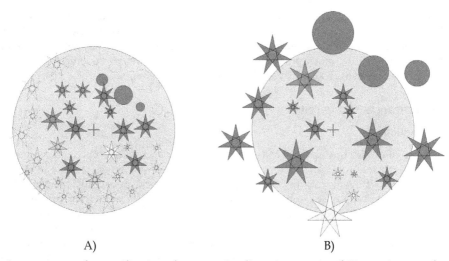

A)   B)

Fig. 6. Assessment and quantification of tree species diversity on a set of 20 trees in case of a dense (A) and released (B) stand. In contrast to sampling on the plots with constant area, the sampled area varies.

Literature survey revealed that remote sensing data have been successfully used for:

1. habitat categorisation and estimation of their changes over large areas (Brotherton, 1983; Cushman & Wallin, 2000 as cited in Humphrey & Watts, 2004)
2. estimation of forest characteristics, e.g. basal area, stand volume, stem density, (Ingram et al., 2005; Maltamo et al., 2006 ; Reese et al., 2003; Tuominen & Haakana, 2005)
3. measuring vegetation (forest) structure (Ingram et al., 2005; Maltamo et al., 2005; Prasad et al., 1998; Wack & Oliveira, 2005)
4. analysis of canopy surface and canopy gaps (Nuske & Nieschulze, 2005)
5. identification of dead standing trees (Butler & Schlaepfer, 2004) and estimation of their amount (Uuttera & Hyppanen, 1998)
6. stratification for ground inventory (Roy & Sanjay-Tomar, 2000; Ghayyas-Ahmad, 2001; Jha et al., 1997) or to increase the precision of estimates (McRoberts et al., 2003, 2005; Olsson et al., 2005)

Nagendra (2001) who evaluated the potential of remote sensing for assessing species diversity distinguished three types of studies:

1. direct mapping of individuals and associations of single species,
2. habitat mapping using remotely sensed data, and prediction of species distribution based on habitat requirements,
3. establishment of direct relationships between spectral radiance values recorded from remote sensors and species distribution patterns recorded from field observations.

Direct mapping is applicable over smaller areas to obtain detailed information on the distribution of certain canopy tree species or associations. Habitat maps appear most capable of providing information on the distributions of large numbers of species in a wider variety of habitat types (Nagendra, 2001).

Turner et al. (2003) recognise two general approaches to the remote sensing of biodiversity. „One is the direct remote sensing of individual organisms, species assemblages, or ecological communities from airborne or satellite sensors. The other approach is the indirect remote sensing of biodiversity through reliance on environmental parameters as proxies" (Turner et al., 2003), that can be clearly identified remotely.

4. Importance of diversity

4.1 Productivity

Experimental relationship between site productivity and biodiversity of community is a widely discussed question in scientific literature. This problem was studied in detail at the end of 1980s (Rozenzweig & Abramsky, 1993). In many cases, this relationship has a humped shape with maximum species diversity at average productivity and minimum at both extremes, i.e. at low and high productivity. This shape was observed both in plant and animal communities. However, no general model that would explain this relationship has been derived yet. The humped shape can be linked with the theory of the limiting factor. On every site each species has a specific productivity threshold. Site factors are limiting for the survival of the species. As the site productivity increases, more and more species exceed their threshold value and hence, can survive in the environment. The decline of diversity with increasing productivity after the peak of the curve is a mystery that has been in the centre of interests of many scientists, who presented several explanations and hypotheses. However, none of them was sufficiently satisfactory. As an example we present two of them.

In the first hypothesis, species diversity is related to micro-site diversity (Rozenzweig & Abramsky, 1993). In theoretical ecology it is a well-known fact that one ecological niche can carry only one species. This theory, also called as "niche theory" says that average sites have more niches than very poor or very rich sites. Hence, we can conclude that they also have higher species diversity. This can be illustrated using three basic site factors: temperature, moisture, nutrients. On poor sites (cold, dry, and nutrient poor), all factors have low values, which results in a unique combination of factors that represent a specific site with very low productivity. Similarly, rich sites (warm, moist, and nutrient rich) are also the result of a unique combination of factors leading to one specific site. However, on average sites, a great number of combinations of site factors exist, while each combination represents a specific niche, which can carry a specific community. According to this theory, site diversity is maximum on average sites, and therefore, species diversity is maximum on average sites.

The second hypothesis is based on the theory of the „right of the limiting factor" (Rozenzweig & Abramsky, 1993). When the site productivity is high, all species have the potential to survive. However, a large number of species on the site leads to a strong competition resulting in the reductions in species number. Hence, low diversity can be caused by the strong competition of the most vital species, which suppress other species.

In forestry applications, this issue is closely related to the production in mixed forest stands, which is becoming an up-to-date theme due to accepting the principles of sustainable and close-to-nature forestry and consequently the transformation of forest management. This management results in greater area of uneven-aged and heterogenous forest stands, which complicates the use of traditional dendrometric models. Some efforts have been made to create models that would enable to determine the volume of several mixture types and forms. From this point of view, tree growth simulators are promising tools that are able to predict the development of mixed forests (Fabrika, 2005; Hasenauer, 1994; Kahn & Pretzsch, 1997; Nagel, 1995; Sterba et al., 1995).

4.2 Stability

Most ecologists agree that species diversity is a good basis for long-term existence of communities, i.e. communities that are composed of only a small number of species are more susceptible to extinction than species-rich communities. Due to this fact, diversity is implicitly linked to stability. However, this theory was disapproved, when May (1973) using Lotka - Volter systems presented that stability decreases with increasing complexity of iterations, i.e. with the increasing value of Simpson diversity index. May´s argument was based on the analysis of system stability through the linearisation of the surrounding balance. In other words, random Lotka-Volter system is stable if it consists of several interconnected species, or if the intensity of connections is low. A lot of important connections lead to system instability. It is still questionable if this is generally valid for all systems. This statement caused wide discussions. Anti-arguments say that in the ecosystem the interconnections are not randomly distributed, but consistently structured, which should cause the increase of stability. A short review of the progress in this field since the work of May (1973) can be found in Sigmund (1995).

In forestry field, stand stability is one of the main principles of sustainable forest management, which was approved in Helsinki Ministerial Conference on the Protection of Forests in Europe. Its importance increases particularly in the last time, which is characterised by more frequent occurrence of large-scale disturbances. Concerning the relationship of stability to biodiversity, Stolina (1996) stated that:

- natural forest ecosystems that are not influenced by anthropogenic activities are characterised by specific species diversity which is adequate to conditions of abiotic environment, because it has resulted from the long-term adaptation process.
- species diversity can be taken as an indicator of forest ecosystem stability;
- not every increase of species diversity measures indicate the increase of stability.

4.3 Naturalness – Diversity indicators of forest naturalness

Both biodiversity and naturalness are frequently used in conservation (Schnitzler et al., 2008), as the criteria for assessing the conservation status of forest ecosystems. Their significance was approved in many international schemes, e.g. both concepts were included in the list of pan-European indicators of sustainable forest management (MCPFE, 2002). The concepts are closely interlinked. For example, the degrees of forest naturalness distinguished within the scope of MCPFE are characterised with regard to biodiversity and its components. In forests undisturbed by man, processes and species composition remain natural to a considerable extent or have been restored. Semi-natural forests can keep certain natural characteristics allowing natural dynamics and biodiversity closer to the original ecosystem. Plantations represent man-made (artificial) forest communities, which are completely distinct from the original ecosystem (MCPFE, 2002).

The objective assessment of forest naturalness presented by several authors (e.g. Bartha et al., 2006; Machado, 2004; Moravčík et al., 2010; Winter et al., 2010) is based on a number of compositional, structural, and functional attributes of biodiversity, such as species composition and structure of different forest layers, occurrence of deadwood, etc. Tree species composition is the most common attribute used for the assessment of forest naturalness (Glončák, 2007; Guarino et al., 2008; Šmídt, 2002; Vladovič, 2003), but recently the amount of deadwood has also gained attention due to the large differences between managed and unmanaged stands. From other structural characteristics, horizontal structure characterised by diameter distribution (Pasierbek et al., 2007), differentiation of vertical and age structure are biodiversity indicators used in the assessment of forest naturalness (Moravčík et al., 2010).

5. Conclusion

Biodiversity is a keystone of ecosystem functioning. Its actual state determines if the ecosystem is sustainable, and hence, if it can fulfil particular functions, or ecosystem services. Since nowadays biodiversity has been receiving much attention worldwide, it is of great importance to understand this term thoroughly and to be able to quantify it mathematically. Various assessment methods and evaluation procedures have been used for the quantification of partial components of biodiversity, which allow users to evaluate and compare ecosystems objectively. In the presented chapter, we reviewed the current state-of-art of plant diversity assessment and examined the relationship of plant diversity to main forestry issues, namely forest management, productivity, stability and naturalness. The review of the available knowledge indicates that for the proper utilisation of biodiversity measures, their values should always refer to the area they represent. The questions about the relationships between biodiversity and forest productivity, stability and consequently management remain open for future research.

6. Acknowledgments

This work was supported by the National Agency for Agriculture Research in Czech Republic under the contract No. QH91077 and the Slovak Research and Development Agency under the contracts No. APVT-27-009304, and APVV-0632-07.

7. References

Aguirre, O.; Kramer, H. & Jiménez, J. (1998). Strukturuntersuchungen in einem Kiefern-Durchforstungsversuch Nordmexikos. *Allgemeine Forst und Jagdzeitung*, Vol. 169, pp. 213-219

Alatalo, R.V. (1981). Problems in the measurement of evenness in ecology. Oikos, Vol. 37, pp. 199-204

Allen, R.B.; Bellingham, P.J. & Wiser, S.K. (2003). Forest biodiversity assessment for reporting conservation performance. *Science-for-Conservation*, Vol. 216, 49 pp.

Austin, M.P.; Pausas, J.G. & Nicholls, A.O. (1996). Patterns of tree species richness in relation to environment in southeastern New South Wales, Australia. *Australian Journal of Ecology*, Vol. 21, No. 2, pp. 154-164

Bachman, S.; Baker, W.J.; Brummitt, N.; Dransfield, J. & Moat, J. (2004). Elevational gradients, area and tropical island diversity: An example from the palms of New Guinea. *Ecography*, 27:297

Bartha, D.; Ódor, P.; Horváth, T.; Tímár, G.; Kenderes, K.; Standovár, T.; Bölöni, J.; Szmorad, F.; Bodonczi, L. & Aszalós, R. (2006). Relationship of Tree Stand Heterogeneity and Forest Naturalness. *Acta Silv. Lign. Hung.*, Vol. 2, pp. 7-22

Battles, J.J. (2001).The effects of forest management on plant diversity in a Sierran conifer forest. *Forest Ecology and Management*, Vol. 146, No. 1-3, pp. 211-222

Belaoussoff, S.; Kevan, P.G.; Murphy, S. & Swanton, C.E. (2003). Assessing tillage disturbance on assemblages of ground beetles (Coleoptera: Carabidae) by using a range of ecological indices. *Biodiversity and Conservation*, Vol. 12, pp. 851-882

Bhattarai, K.R. & Vetaas, O.R. (2003). Variation in plant species richness of different life forms along a subtropical elevation gradient in the Himalayas, east Nepal. *Glob. Ecol. Biogeogr.*, Vol. 12, pp. 327-340

Bhattarai, K.R.; Vetaas, O.R. & Grytnes, J.A. (2004). Fern species richness along a central Himalayan elevational gradient. *Nepal. J. Biogeogr.*, Vol. 31, pp. 389-400

Brillouin, L. (1956). *Science and Information Theory*. Academic Press, New York

Bruciamacchie, M. (1996). *Comparison between indices of species diversity*. Munich, Vol. 3/96, 14p.

Burns, B.R. (1995). Environmental correlates of species richness at Waipoua Forest Sanctuary, New Zealand. *New Zealand Journal of Ecology*, Vol. 19, No. 2, pp. 153-162

Butler, R. & Schlaepfer, R. (2004). Spruce snag quantification by coupling colour infrared aerial photos and a GIS. *Forest-Ecology-and-Management*, Vol. 195, No. 3, pp. 325-339

Christensen, M.; Heilmann, C.J.; Walleyn, R. & Adamcik, S. (2004). Wood-inhabiting Fungi as Indicators of Nature Value in European Beech Forests. *EFI Proceedings*, Vol. 51, pp. 229-238

Clark, P.J. & Evans, F.C. (1954). Distance to nearest neighbour as a measure of spatian relationship in populations. *Ecology*, Vol. 35, p. 445-453

Çolak, A.H.; Rotherham, I.D. & Çalikoglu, M. (2003). Combining 'Naturalness Concepts' with Close-to-Nature Silviculture. *Forstw. Cbl.*, Vol. 122, pp. 421-431

Cole, D.N.; Yung, L.; Zavaleta, E.S.; Aplet, G.H.; Chapin, F.S.I.; Graber, D.M.; Higgs, E.S.; Hobbs, R.J.; Landres, P.B.; Millar, C.I.; Parsons, D.J.; Randall, J.M.; Stephenson, N.L.; Tonnessen, K.A.; White, P.S. & Woodley, S. (2008). Naturalness and Beyond: Protected Area Stewardship in an Era of Global Environmental Change. *The George Wright Forum*, Vol. 25, No. 1, pp. 36-55

Cooley, W.W. & Lohnes, P.R. (1971). *Multivariate data analysis*. Wiley, New York

Corona P. & Marchetti, M. (2007). Outlining multi-purpose forest inventories to assess the ecosystem approach in forestry. *Plant Biosystems*, Vol. 141, No. 2, pp. 243-251

Cox, F. (1971). Dichtebestimung und Strukturanalyse von Pflanzenpopulationen mit Hilfe von Abstandsmessungen. *Mitt. Bundesforschungsanstalt Forst- und Holzwirtschaft Reinbeck*, No. 87, 161 pp.

Eriksson, S. & Hammer, M. (2006).The challenge of combining timber production and biodiverzity conservation for longterm ecosystem functioning—a case study of Swedish boreal forestry. *Forest Ecology and Management*, Vol. 237, No. 1-3, pp. 208-217

Estreguil, C.; Vogt, P.; Cerruti, M. & Maggi, M. (2004). JRC Contribution to Reporting Needs of EC Nature and Forest Policies. *EFI Proceedings*, Vol. 51, pp. 91-104

Fabrika, M. (2005). *Simulátor biodynamiky lesa SIBYLA*. Koncepcia, konštrukcia a programové riešenie. Habilitačná práca. Technická univerzita vo Zvolene, 238 p.

Fager, E.W. (1972). Diversity: a sampling study. *Am. Natur.*, Vol. 106, pp. 293-310

Ferrari C.; Pezzi, G.; Diani, L.; Corazza, M. (2008). Evaluating landscape quality with vegetation naturalness maps: an index and some inferences. *Appl. Veg. Sci.*, Vol. 11, pp. 243-250

Ferris, R. & Humphrey, J.W. (1999). A review of potential biodiversity indicators for application in British forests. *Forestry-Oxford*, Vol. 72, No. 4, pp. 313-328

Ferris-Kaan, R.; Peace, A.J. & Humphrey, J.W. (1998). Assessing structural diversity in managed forests, In: *Assessment of Biodiversity for Improved Forest Planning*, P. Bachmann; M. Köhl & R. Päivinen, (Ed.), Proceedings of the Conference on Assessment of Biodiversity for Improved Forest Planning, 7 - 11 October 1996, held in Monte Verita, Switzerland, Kluwer Academic Publishers, Dordrecht

Foody, G.M, Cutler, M.E.J. (2003). Tree biodiversity in protected and logged Bornean tropical rain forests and its measurement by satellite remote sensing. *Journal-of-Biogeography*, Vol. 30, No. 7, pp. 1053-1066

Franklin J.F. (1998). The Natural, the Clearcut, and the Future, In: *Proceedings of a workshop on Structure, Process, and Diversity in Successional Forests of Coastal British Columbia*, J.A. Trofymow & A. MacKinnon (Ed.), February 17-19, 1998, Victoria, British Columbia, Northwest Science, Vol. 72, No. 2, pp. 134-138

Frelich, L.E.; Sugita, S.; Reich, P.B.; Davis, M.B. & Friedman, S.K. (1998). Neighbourhood effects in forests: implications for within stand patch structure. *Journal of Ecology*, Vol. 86, pp. 149-161

Füldner, K. (1995). Zur Strukturbeschreibung in Mischbeständen. *Forstarchiv*, 66, p. 235-240

Gadow, K. & Hui, G. (1999). *Modelling forest development*. Kluwer Academic Publishers, Dordrecht, 213 p.

Gadow, K. (1993). Zur Bestandesbeschreibung in der Forsteinrichtung. *Forst und Holz*, Vol. 21, pp. 602-606

Gadow, K. (1999). Waldstruktur und Diversität. *AFJZ*, Vol. 170, No. 7, pp.117-121

Gadow, K.; Hui, G.Y. & Albert,M. (1998). Das Winkelmaß – ein Strukturparameter zur Beschreibung der Individualverteilung in Waldbeständen. *Centralblatt für das gesamte Forstwesen*, Vol. 115, pp. 1-10

Ghayyas, A. (2001). Identifying priority forest areas in the Salt Range of Pakistan for biodiversity conservation planning using remote sensing and GIS. *Pakistan-Journal-of-Forestry*, Vol. 51, No. 1, pp. 21-40

Gilliam, F.S. & Roberts, M.S. (1995). Forest management and plant diversity. Ecological Society of America, *Ecological Applications*, Vol. 5, No. 4, pp. 911-912

Gleichmar, W. & Gerold, D. (1998). Indizes zur Charakterisierung der horizontalen Baumverteilung. *Forstw. Cbl.*, Vol. 117, pp. 69-80

Glončák, P. (2007). Assessment of forest stands naturalness on the base of phytosociological units (example from the protection zone of Badínsky prales), In: *Geobiocenologie a její aplikace*, V. Hrubá & J. Štykar (Ed.), Geobiocenologické spisy, Vol. 11, pp. 39-46

Grytnes, J.A. & Vetaas, O.R. (2002). Species richness and altitude: A comparison between null models and interpolated plant species richness along the Himalayan altitudinal gradient, *Nepal. The Am. Naturalist*, Vol. 159, pp. 294-304

Guarino, C.; Santoro, S. & De Simone, L. (2008). Assessment of vegetation and naturalness: a study case in Southern Italy, *IForest*, 1, pp. 114-121

Götmark, F. (1992). Naturalness as an Evaluation Criterion in Nature Conservation: A response to Anderson, *Conservation Biology*, Vol. 6, No. 3, pp. 455-458

Hansen, A. & Rotella, J. (1999). The macro approach, managing forest landscapes: Abiotic factors, In: *Maintaining biodiversity in forest ecosystems*, M.L.Jr. Hunter, (Ed.), Cambridge University Press, 665 p., pp. 161-209

Harrison, I.; Laverty, M. & Sterling, E. (2004). Species Diversity, *Connexions module: m12174*, 05.08.2011, Available from http://cnx.org/content/m12174/latest/

Hasenauer, H. (1994). *Ein Einzelbaumwachstumssimulator für ungleichatrige Fichten-Kiefern- und Buchen-Fichtenmischbestände*. Forstl. Schriftenreihe Univ. f. Bodenkultur., 8. Österr. Ges. f. Waldökosystemforschung und experimentelle Baumforschung an der Univ. f. Bodenkultur, Wien. 152 p.

Haußmann, T. & Fischer, R. (2004). The Forest Monitoring Programme of ICP Forests – A Contribution to biodiversity monitoring. *EFI Proceedings*, Vol. 51, pp. 415-419

Hédl, R. & Kopecký, M. (2006). Hospodaření v lese a biodiverzita. *Živa*, 3

Hédl, R. (2006). *Stav lesů v ČR z ekologické perspektivy*. 05.08.2011, Available from http://sweb.cz/diskuse.lesy/

Heer, M.; De Kapos, V.; Miles, L. & Brink, B.T. (2004). Biodiversity Trends and Threats in Europe – Can we apply a generic biodiversity indicator to forests? *EFI Proceedings* Vol. 51, pp. 15-26

Heip, C. (1974). A new index measuring evenness. *Journal of Marine Biological Association*, Vol. 54, pp. 555-557

Hill, M.O. (1973). Diversity and Evenness: a unifying notation and its consequences. *Ecology*, Vol. 54, No. 2, pp. 427-432

Holdridge, L.R. (1967). *Life zone ecology*. Tropical Science Center, San JoseÂ, Costa Rica.

Holopainen, M.; Talvitie, M. & Leino, O. (2005). Forest biodiversity inventory by means of digital aerial photographs and laser scanner data. *Schriften aus der Forstlichen Fakultät der Universität Göttingen und der Niedersachsischen Forstlichen Versuchsanstalt*, Vol. 138, pp. 340-348

Hubálek, Z. (2000). Measures of species diversity in ecology: an evaluation. *Folia Zool.*, Vol. 49, No. 4, pp. 241-260

Huberty, C.J. (1994). *Applied discriminant analysis*. Wiley and Sons, New York

Humphrey, J.W. & Watts,K. (2004). Biodiversity indicators for UK managed forests: development and implementation at different spatial scales. *EFI-Proceedings*, Vol. 51, pp. 79-89

Hurlbert, S.H. (1971). The nonconcept of species diversity: a critique and alternative parameters. *Ecology*, Vol. 52, pp. 577-586

Hutcheson, K. (1970). A test for comparing diversities based on the Shannon formula. *J. Theor. Biol.*, Vol. 29, pp. 151–154

Hyyppa, J.; Hyyppa, H.; Inkinen, M.; Engdahl, M.; Linko, S. & Zhu Y. (2000). Accuracy comparison of various remote sensing data sources in the retrieval of forest stand attributes. *Forest Ecology and Management*, Vol. 128, No. 1/2, pp. 109-120

Ingram, J.C.; Dawson, T.P. & Whittaker, R.J. (2005). Mapping tropical forest structure in southeastern Madagascar using remote sensing and artificial neural networks. *Remote Sensing of Environment*, Vol. 94, No. 4, pp. 491-507

Innes, J.L. & Koch, B. (1998). Forest biodiversity and its assessment by remote sensing. *Global Ecology and Biogeography Letters*, Vol. 7, No. 6, pp. 397-419

Jaehne, S. & Dohrenbusch, A. (1997). Ein Verfahren zur Beurteilung der Bestandesdiversität. *Forstw. Cbl.*, Vol. 116, pp. 333-345

Jha, C.S.; Udayalakshmi,V. & Dutt, C.B.S. (1997). Pattern diversity assessment using remotely sensed data in the Western Ghats of India. *Tropical-Ecology*, Vol. 38, No. 2, pp. 273-283

Johnson, F.L. (1986). Woody vegetation of Southeastern LeFlore County, Oklahoma, in relation to topography. *Proc. Oklahoma Acad. Sci.*, Vol. 66, pp. 1-6

Kaennel, M. (1998). BIODIVERSITY: a Diversity in Definition, In: *Assessment of Biodiversity for Improved Forest Planning*, P. Bachmann; M. Köhl & R. Päivinen, (Ed.), Proceedings of the Conference on Assessment of Biodiversity for Improved Forest Planning, 7 - 11 October 1996, held in Monte Verita, Switzerland, Kluwer Academic Publishers, Dordrecht, pp. 71-82

Kahn, M. & Pretzsch, H. (1997). Das Wuchsmodell SILVA-Parametrisierung der Version 2.1 für Rein- und Mischbestände aus Fichte und Buche. *AFJZ*, Vol. 168, No. 6-7, pp. 115-123

Kempton, R.A. & Taylor, L.R. (1976). Models and statistics for species diversity. *Nature*, Vol. 262, pp. 818–820

Kint, V.; Lust, N.; Ferris, R. & Olsthoorn, A.F.M. (2000). *Quantification of Forest Stand Structure Applied to Scots Pine (Pinus sylvestris L.)*. Forests. Invest. Agr.: Sist. Recur. For.: Fuera de Serie n.° 1, 17 p.

Koch, B. & Ivits, E. (2004). Reults from the Project BioAssess – relation between remote sensing and terrestrial derived biodiversity indicators. *EFI Proceedings*, Vol. 51, pp. 315-332

Kozak, J.; Estreguil, C. & Vogt, P. (2007). Forest cover and pattern changes in the Carpathians over the last decades. *European Journal of Forest Research*, Vol. 126, pp. 77-90

Krebs, C.J. (1989). *Ecological methodology*. Harper and Row, New York, 471 p.

Larsson, T.B. (2001). Biodiversity evaluation tools for European forest. *Ecological Bulletin*, Vol. 50, pp. 1-237

Laxmi, G.; Anshuman, T. & Jha, C.S. (2005). Forest fragmentation impacts on phytodiversity - an analysis using remote sensing and GIS. *Current-Science*, Vol. 88, No. 8, pp. 1264-1274

Lefsky, M.A.; Cohen, W.B. & Spies,T.A. (2001). An evaluation of alternate remote sensing products for forest inventory, monitoring, and mapping of Douglas-fir forests in western Oregon. *Canadian Journal of Forest Research*, Vol. 31, No. 1, pp. 78-87

Lexer, M.J.; Lexer, W. & Hasenauer, H. (2000). The Use of Forest Models for Biodiversity Assessments at the Stand Level. *Invest. Agr.: Sist. Recur. For.: Fuera de Serie*, Vol. 1, pp. 297-316

Lindenmayer, D.B. (1999). Future directions for biodiversity conservation in managed forests: indicator species, impact studies and monitoring programs. *Forest Ecology and Management*, Vol. 115, No. 2/3, pp. 277-287

Lloyd, M. & Ghelardi, R.J. (1964). A table for calculating the "equitability" component of species diverzity. *J. Anim. Ecology*, Vol. 33, pp. 217-225

Ludwig, J.A. & Reynolds, J.F. (1988). *Statistical Ecology a primer on methods and computing.* John Willey & Sons, 337 p.

Lübbers, P. (1999). *Diversitätsindizes und Stichprobenverfahren.* Universität Freiburg, 10p.

Lähde, E.; Laiho, O.; Norokorpi, Y. & Saksa, T. (1999). Stand structure as the basis of diversity index. *Forest Ecology and Management*, Vol. 115, pp. 213-220

Machado, A. (2004). An index of naturalness. *Journal of Nature Conservation*, Vol. 12, pp.95-110

Maltamo, M.; Malinen, J.; Packalen, P.; Suvanto, A. & Kangas, J. (2006). Nonparametric estimation of stem volume using airborne laser scanning, aerial photography, and stand-register data. *Canadian Journal of Forest Research*, Vol. 36, No. 2, pp. 426-436

Margalef, R. (1958). Information theory in ecology. *General Systematics*, Vol. 3, pp. 36-71

May, R.M. (1973). *Stability and Complexity in Model Ecosystems.* Princeton University Press.

McIntosh, R.P. (1967). An index of diversity and the relation of certain concepts to diversity. *Ecology*, Vol. 48, pp. 392-404

MCPFE (2002). Improved Pan-European indicators for sustainable forest management as adopted by the MCPFE Expert Level Meeting 2002. 27.04.2011, Available from http://www.mcpfe.org/system/files/u1/Vienna_Improved_Indicators.pdf

McRoberts, R.E.; Gormanson, D.D. & Hansen, M.H. (2005). Can a land cover change map be used to increase the precision of forest inventory change estimates? *Schriften-aus-der-Forstlichen-Fakultat-der-Universitat-Gottingen-und-der-Niedersachsischen-Forstlichen-Versuchsanstalt*, Vol. 138, pp. 74-82

McRoberts, R.E.; Nelson, M.D. & Wendt, D.G. (2003). Stratified estimates of forest area using the k-Nearest Neighbors technique and satellite imagery. *General-Technical-Report-North-Central-Research-Station,-USDA-Forest-Service*, (NC-230), pp. 80-86

Meersschaut, D.V.D. & Vandekerkhove, K. (1998). Development of a stand-scale forest biodiversity index based on the State Forest Inventory. *Integrated Tools Proceedings Boise*, Idaho, USA, August 16-20., pp. 340-349

Menhinick, C.F. (1964). A comparison of some species – individuals diversity indices applied to samples of field insects. *Ecology*, Vol. 45, pp. 859-861

Merganič, J. & Šmelko, Š. (2004). Quantification of tree species diversity in forest stands Model BIODIVERSS. *European Journal of Forest Research*, Vol. 2, No. 123, pp. 157-165

Merganič, J.; Quednau, H.D. & Šmelko, Š. (2004). Influence of Morphometrical Characteristics of Georelief on Species Diversity of Forest Ecosystems and its Regionalisation. *European Journal of Forest Research*, Vol. 1, No. 123, pp. 75-85

Misir, N.; Misir, M.; Karahalil, U. & Yavuz, H. (2007). Characterization of soil erosion and its implication to forest management. *Journal of Environmental Biology*, Vol. 28, pp. 185-191

Molinari, J. (1989). A calibrated index for the measurement of evenness. *Oikos*, Vol. 56, pp. 319-326

Moravčík, M.; Sarvašová, Z.; Merganič, J. & Schwarz, M. (2010). Forest naturalness – decision support in management of forest ecosystems. *Environmental Management*, Vol. 46, No. 6, pp. 908-919

Nagaraja, B.C.; Somashekar, R.K. & Raj, M.B. (2005). Tree species diversity and composition in logged and unlogged rainforest of Kudremukh National Park, south India. *Journal of Environmental Biology*, Vol. 26, pp. 627-634

Nagel, J. (1995). BWERT: Programm zur Bestandesbewertung und zur Prognose der Bestandesentwickung. *DFFA, Sektion Ertragskunde, Jahrestagung Joachimsthal*, pp. 184-198

Nagendra, H. (2001). Using remote sensing to assess biodiversity. *International Journal of Remote Sensing*, Vol. 22, No. 12, pp. 2377-2400

Neumann, M. & Starlinger, F. (2001). The significance of different indices for stand structure and diversity in forests. *Forest Ecology and Management*, Vol. 145, pp. 91-106

Newmaster, S.G. (2007). Effects of forest floor disturbances by mechanical site preparation on floristic diversity in a central Ontario clearcut. *Forest Ecology and Management*, Vol. 246, pp. 196-207

Noss, R.F. (1990). Indicators for monitoring biodiversity: a hierarchical approach. *Conservation Biology*, Vol. 4, No. 4, pp. 355-364

Nunes, P.A.L.D.; Bergh J.C.J.M. & Nijkamp, P. (2000). Ecological− Economic Analysis and Valuation of Biodiversity. *Tinbergen Institute Discussion Paper 2000-100/3*. 05.02.2011, Available from http ://www.tinbergen.nl

Nuske, R.S. & Nieschulze, J. (2005). Remotely sensed digital height models and GIS for monitoring and modeling ecological characteristics of forest stands. *Schriften-aus-der-Forstlichen-Fakultat-der-Universitat-Gottingen-und-der-Niedersachsischen-Forstlichen-Versuchsanstalt*, Vol. 138, pp. 83-92

Ode, Å.; Fry, G.; Tveit, M.S.; Messager, P. & Miller, D. (2009). Indicators of perceived naturalness as drivers of landscape preference. *J. Environ. Manage*, Vol. 90, pp. 375

Ohmann, J.L. & Spies, T.A. (1998). Regional gradient analysis and spatial pattern of woody plant communities of Oregon forests. *Ecological Monographs*, Vol. 68, No. 2, pp. 151-182

Olsson, H.; Nilsson, M.; Hagner, O.; Reese, H.; Pahlen, T.G. & Persson, A. (2005). Operational use of Landsat-/SPOT-type satellite data among Swedish forest authorities. *Schriften-aus-der-Forstlichen-Fakultat-der-Universitat-Gottingen-und-der-Niedersachsischen-Forstlichen-Versuchsanstalt*, Vol. 138, pp. 195-203

Ozcelik, R.; Gul, A.U.; Merganič, J. & Merganičová, K. (2008). Tree species diversity and its relationship to stand parameters and geomorphology features in the eastern Black sea region forests of Turkey. *Journal of Environmental Biology*, Vol. 29, No. 3, pp. 291-298, ISSN: 0254-8704

Ozdemir, I.; Asan, U.; Koch, B.; Yesil, A.; Ozkan, U.Y. & Hemphill, S. (2005). Comparison of Quickbird-2 and Landsat-7 ETM+ data for mapping of vegetation cover in Fethiye-Kumluova coastal dune in the Mediterranean region of Turkey. *Fresenius-Environmental-Bulletin*, Vol. 14, No. 9, pp. 823-831

Palmer, M.W.; Clark, D.B. & Clark, D.A. (2000). Is the number of tree species in small tropical forest plots nonrandom? *Community Ecology*, Vol. 1, pp. 95-101

Pasierbek T.; Holeksa J.; Wilczek Z. & Żywiec M. (2007). Why the amount of dead wood in Polish forest reserves is so small? *Nature Conservation*, Vol. 64, pp. 65-71

Pausas, J.G. & Saez, L. (2000). Pteridophyte richness in the NE Iberian Peninsula: Biogeographic patterns. *Plant Ecology*, Vol. 148, pp. 195-205

Pausas, J.G.; Carreras, J.; Ferre, A. & Font, X. (2003). Coarse-scale plant species richness in relation to environmental heterogeneity. *J. Veg. Sci.*, Vol. 14, pp. 661-668

Peet, R.K. (1974). The measurement of species diversity. *Ann. Rev. Ec. Sys.*, Vol. 5, pp. 285-307

Pielou, E.C. (1959). The use of point to plant distances in the study of the pattern of plant populations. *J. Ecol.*, Vol. 47, pp. 607-613

Pielou, E.C. (1969). *An introduction to mathematical ecology*. Wiley, New York, 280 p.

Pielou, E.C. (1975). *Ecological Diversity*. Wiley, New York

Pielou, E.C. (1977). *Mathematical Ecology*. Willey, New York

Poleno, Z. (1997). *Trvale udržitelné obhospodařování lesů*. MZe ČR, Praha.

Pommerening, A. (2002). Approaches to quantifying forest structures. *Forestry-Oxford*, Vol. 75, No. 3, pp. 305-324

Prasad, V.K.; Rajagopal, T.; Yogesh, K. & Badarinath, K.V.S. (1998). Biodiversity studies using spectral and spatial information from IRS-1C LISS-III satellite data. *Journal of Ecobiology*, Vol. 10, No. 3, pp. 179-184

Pretzsch, H. (1996). Strukturvielfalt als Ergebnis Waldbaulichen Handels. *AFJZ*, Vol. 167, p. 213-221

Pretzsch, H. (1998). Structural Diversity as a result of silvicultural operations. *Lestnictví-Forestry*, Vol. 44, No. 10, p. 429-439

Prevosto, B. (2011). Effects of different site preparation treatments on species diversity, composition, and plant traits in Pinus halepensis woodlands. *Plant Ecology*, Vol. 212, No. 4, pp. 627-638

Ramovs, B.V.; Roberts, M.R. (2005). Response of plant functional groups within plantations and naturally regenerated forests in southern Brunswick. NRC Research Press, *Canadian Journal of Forest Ressearch*, Vol. 35, No. 6, pp. 1261-1276

Rautjärvi, N.; Luquea, S. & Tomppo,E. (2005). Mapping spatial patterns from National Forest Inventory data: a regional conservation planning tool. *Schriften-aus-der-Forstlichen-Fakultat-der-Universitat-Gottingen-und-der-Niedersachsischen-Forstlichen-Versuchsanstalt*, Vol. 138, pp. 293-302

Ravindranath, N.H.M. (2006). Community forestry initiatives in Southeast Asia: a review of ecological impacts. *International Journal of Environment and Sustainable Development*, Vol. 5, No. 1, pp. 1-11

Reese, H.; Nilsson, M.; Pahlén, T.G.; Hagner, O.; Joyce, S.; Tingelöf, U.; Egberth, M. & Olsson, H. (2003). Countrywide Estimates of Forest Variables Using Satellite Data and Field Data from the National Forest Inventory. *Ambio*, Vol. 32, No. 8, pp. 542-548

Rego, F.; Godinho F.P.; Uva, J.S. & Cunha, J. (2004). Combination of structural and compositional factors for describing forest types using National Forest Inventory data. *EFI-Proceedings*, Vol. 51, pp. 153-162

Reichholf, J. (1999). *Les. Ekologie středoevropských lesů*. Ikar, Praha. ISBN 80-7202-494-9

Rosenzweig, M.L. (1995). *Species diversity in space and time*. Cambridge University Press, 665 p.

Roy, P.S. & Sanjay, T. (2000). Biodiversity characterization at landscape level using geospatial modelling technique. *Biological Conservation*, Vol. 95, No. 1, pp. 95-109

Rozenzweig, M.L. & Abramsky, Z. (1993). How are diversity and productivity related? In: *Species diversity in ecological Communities. Historical and geographical perspectives*, Ricklefs, E. & Schluter, D. (Ed.)

Sanders, H.L. (1968). Marine benthic diversity: a comparative study. *Am. Natur.*, Vol. 102, pp. 243-282

Schardt, M.; Grancia, K.; Hischmugl, M.; Luckel, H.W. & Klaushofer, F. (2005). Mapping Protection Forests in the Province of Salzburg Using Remote Sensing. *Schriften-aus-der-Forstlichen-Fakultat-der-Universitat-Gottingen-und-der-Niedersachsischen-Forstlichen-Versuchsanstalt*, Vol. 138, pp. 204-213

Schnitzler A.; Génot J.C.; Wintz M. & Hale B.W. (2008). Naturalness and conservation in France. *J. Agr. Environ. Ethic.*, Vol. 21, pp. 423-436

Sepp, T. & Liira, J. (2009). Factors influencing the species composition and richness of herb layer in old boreo-nemoral forests. *Forestry Studies/Metsanduslikud Uurimused*, Vol. 50, pp. 23-41

Shannon, C. & Weaver, W. (1949). *The Mathematical Theory of Communication*. University of Illinois Press. Urbana. Illinois

Sheldon, A.L. (1969). Equitability indices: dependence on the species count. *Ecology*, Vol. 50, pp. 466-467

Sigmund, K. (1995). Darwins circle of complexity: assembling ecological communities. *Complexity*, Vol. 1, pp. 40-44

Simpson, E.H. (1949). Measurement of diversity. *Nature*, Vol. 163, pp. 688

Šmelko, Š. & Fabrika, M. (2007). Evaluation of qualitative attributes of forest ecosystems by means of numerical quantifiers. *J. For. Sci.*, Vol. 53, No. 12, pp. 529-537

Šmelko, Š. (1997). Veľkoplošná variabilita porastových veličín v lesoch Slovenska a faktory, ktoré ju ovplyvňujú. *Acta facultatis forestalis*, Zvolen, XXXIX, pp. 131-143

Šmídt, J. (2002). Method of the assessment of forest natzralness in national park Muránska planina, In: *Výskum a ochrana prírody Muránskej planiny*, Uhrin, M. (Ed), 3. Správa NP Muránska planina, Bratislava & Revúca, pp. 119-123

Smola, M. (2008). *Hospodaření v lesích na principech trvalosti a vyrovnanosti*. Pracovní metodika pro privátní poradce v lesnictví. UHÚL, Brandýs nad Labem

Sobek, E.A & Zak, J.C. (2003). The Soil FungiLog procedure: method and analytical approaches toward understanding fungal functional diversity. *Mycologia*, Vol. 95, No. 4, pp. 590-602

Spies, T. & Turner, M. (1999). The macro approach, managing forest landscapes: Dynamic forest mosaics, In: *Maintaining biodiversity in forest ecosystems*, Hunter, M.L.Jr. (Ed.), Cambridge University Press, 665 p., pp. 95-160

StatSoft (2004). *STATISTICA for Windows*. Tulsa, OK http://www.statsoft.com

Sterba, H.; Moser, M. & Monserud, R.A. (1995). PROGNAUS - Ein Waldwachstum-simulator für Rein- und Mischbestände. *ÖFZ*, Vol. 106, 5, pp. 19-20

Stofer, S. (2006). Working Report ForestBIOTA - Epiphytic Lichen Monitoring. 05.02.2011, Available from http://www.forestbiota.org/

Stokland, J.N.; Eriksen, R.; Tomter, S.M.; Korhonen, K.; Tomppo, E.; Rajaniemi, S.; Söderberg, U.; Toet, H. & Riis Nielsen, T. (2003). *Forest biodiversity indicators in the Nordic countries. Status based on national forest inventories*. TemaNord, Copenhagen. 108 p.

Stolina, M. (1996). Biodiverzita, odolnostný potenciál a ochrana lesných ekosystémov, In: *Biodiverzita z aspektu ochrany lesa a poľovníctva*, Hlaváč,P. (Ed.), Zborník referátov z konferencie, TU Zvolen, pp. 13-19

Strand, L. (1953). Mal for fordelingen av individer over et omrade. *Det Norske Skogforsoksvesen*, Vol. 42, pp. 191-207

Söderberg, U. & Fridman, J. (1998). Monitoring of Forest Biodiversity from Forest resource Inentory Data. *Assessment of biodiversity for improved forest planning Proceedings of the conference on assessment of biodiversity of improved forest planning, 7-11 October 1996, Monte Verita, Switzerland*. pp. 233-240

Terradas, J.; Salvador, R.; Vayreda, J. & Loret, F. (2004). Maximal species richness: An empirical approach for evaluating woody plant forest biodiversity. *Forest Ecology and Management*, Vol. 189, pp. 241-249

Thompson, M.W. & Whitehead, K. (1992). An overview of remote sensing in forestry and related activities: its potential application in South Africa. *South-African-Forestry-Journal*, Vol. 161, pp. 59-68

Tomppo, E. (1986). *Models and methods for analysing spatial patterns of trees*. Communicationes Instituti Forestalis Fenniae, Vol. 138, 65 p.

Tuominen, S. & Haakana, M. (2005). Landsat TM imagery and high altitude aerial photographs in estimation of forest characteristics. *Silva-Fennica*, Vol. 39, No. 4, pp. 573-584

Turner, W.; Spector, S.; Gardiner, N.; Fladeland, M.; Sterling, E. & Steininger, M. (2003). Remote sensing for biodiversity science and conservation. *TRENDS in Ecology and Evolution*, Vol. 18, No. 6, pp. 306-314

Ucler, A.O.; Yucesan, Z.; Demirci, A.; Yavuz, H. & Oktan, E. (2007). Natural tree collectives of pure oriental spruce [Picea orientalis (L.) Link] on mountain forests in Turkey. *J. Environ Biol.*, Vol. 28, pp. 295-302

Uuttera, J. & Hyppanen, H. (1998). Determination of potential key-biotope areas in managed forests of Finland using existing inventory data and digital aerial photographs. *Forest-and-Landscape-Research*, Vol. 1, No. 5, pp. 415-429

Vladovič, J. (2003). *Regional principles of evaluation of tree species composition and ecological stability of Slovak forests*. Lesnícke štúdie 57, Príroda, Bratislava, 160 p.

Wack, R. & Oliveira, T. (2005). Analysis of vertical vegetation structures for fire management by means of airborne laser scanning. *Schriften-aus-der-Forstlichen-Fakultat-der-Universitat-Gottingen-und-der-Niedersachsischen-Forstlichen-Versuchsanstalt*. Vol. 138, pp. 127-135.

Wang, S. & Chen, H.Y.H. (2010). Diversity of northern plantations peaks at intermediate management intensity. *Forest Ecology and Management*, Vol. 259, No. 3, pp. 360–366

Wezyk, P.; Tracz, W. & Guzik, M. (2005). Evaluation of landscape structure changes in Tatra Mountains (Poland) based on 4D GIS analysis. *Schriften-aus-der-Forstlichen-Fakultat-der-Universitat-Gottingen-und-der-Niedersachsischen-Forstlichen-Versuchsanstalt*. Vol. 138, pp. 224-232

Williams, K. (2002). Beliefs about natural forest systems. *Aust. Forestry*, Vol. 65, No. 2, pp. 81-86

Winter, S.; Fischer, H.S. & Fischer, A. (2010). Relative Quantitative Reference Approach for Naturalness Assessments of forests. *Forest Ecology and Management*, Vol. 259, pp. 1624–1632

Zenner, E.K. & Hibbs, D.E. (2000). A new method for modeling the heterogeneity of forest structure. *Forest Ecology and Management*, Vol. 129, pp. 75-87

Zenner, E.K. (1999). *Eine neue Methode zur Untersuchung der Dreidimensionalität in Waldbeständen*. Universität Freiburg, 11 p.

3

Advances in Molecular Diversity of Arbuscular Mycorrhizal Fungi (Phylum Glomeromycota) in Forest Ecosystems

Camila Maistro Patreze, Milene Moreira and Siu Mui Tsai
UNIRIO- Universidade Federal do Estado do Rio de Janeiro, APTA – Polo Centro Sul,
Laboratório de Biologia Celular e Molecular, CENA – Universidade de São Paulo
Brazil

1. Introduction

Glomeromycota is a fungal phylum scientifically recognized in 2001 as monophyletic group which probably diverged from the same common ancestor as the Ascomycota and Basidiomycota (Schüssler et al., 2001). Glomeromycota comprises arbuscular mycorrhizal fungi (AMF), important simbiont of land plants and the endocytobiotic fungus, *Geosiphon pyriformis*. Despite of fungi worldwide distribution, relatively few species have been described. The AM fungi are an interesting case: like other fungi they are distributed organisms with apparently widespread mycelial networks, but no sexual stage has been identified in any species in the phylum, they cannot easily be observed or located in situ as they have no conspicuous above-ground fruiting body, and they cannot yet be grown in axenic culture in the absence of plant roots. From 1,5 million microbial fungal species in soils worldwide estimated (Torsvik et al., 1990), about 214 currently described species are glomeromycetes within a universe of 97,000 species of fungal kingdom described so far.

Mycorrhizas are ubiquitous in terrestrial ecosystems. Around 80% of plant species that have been studied form the symbiosis (Wang & Qiu, 2006). There are several different types of mycorrhiza, in which different plant and fungal taxa are involved, but the arbuscular mycorrhizas (AMs) form a monophyletic group of obligate plant symbiotic fungi belonging to the Phylum Glomeromycota and around two-thirds of plant species, is both the most widespread and ancestral: all modern plants have ancestors that formed AM symbioses (Helgason & Fitter, 2009).

The AM fungi are an interesting case: like other fungi they are distributed organisms with apparently widespread mycelial networks, but no sexual stage has been identified in any species in the phylum, they cannot easily be observed or located in situ as they have no conspicuous above-ground fruiting body, and they cannot yet be grown in axenic culture in the absence of plant roots. Initially, the AM species were described only based on asexual spore's morphology. However, new findings showed problems related to dimorphic spores in some species, as for example, spores with morphs from *Glomus intraradices*, i.e., spore with wall simple and Scutellospora spp. morphs, i.e., spores containing several inner membranous walls and develop complex germination shields at the same species.

Furthermore, the scientific community remains discussing several morphologic characters and its limitations to the species identification. Studies based on rDNA sequence have often confirmed the morphologically defined species and the molecular data have erected new genera and families, revealing a considerable unknown AM diversity. To identify fungi from various substrates and in different life stages to species level, the most frequently sequenced region is the internal transcribed spacer (ITS) region of the nuclear ribosomal DNA, however other regions are informative as 18S rRNA, small subunit ribosomal of rRNA (SSU), large subunit ribosomal of rRNA (LSU). Thus, molecular methods are potential tools to cover the diversity of this group.

It is well known that mycorrhizal associations take up mineral nutrients from the soil and exchange nutritional elements with plants for photosynthetically fixed carbon (Smith & Read, 2008). The main nutrient is phosphorous (P) and AMF structures assimilates P from lower concentrations in the soil at which normal plant roots fail (Jefferies et al., 2003). Moreover, these fungi can uptake K, Zn, Fe, Cu, Mg and Ca. The mechanism of absorption is based on major surface area of roots of plants and exploring soil by extraradical hyphae beyond the root hair. The element is then biochemically transformed at external hyphae and passed to the arbuscules for being ultimately transferred to the host plant (Azcon-Aguilar & Barea, 1996).

The plant hosts of AMF are mostly angiosperms, some gymnosperms, pteridophytes, lycopods and mosses (Smith & Read, 1997) which comprises over 80% of all terrestrial plant species. The AM symbiosis is associated with a range of additional benefits for the plant including the acquisition of other mineral nutrients, such as nitrogen and resistance to a variety of stresses (Kaapor et al., 2008) or pathogens (Rabie, 1998), soil aggregation, and carbon sequestration (van der Heijden et al., 2008). AMF can mitigate the effects of extreme variations in temperature, pH and water stress (Michelsen & Rosendahl, 1990; Siqueira, 1994; Augé 2001). For water stress, recent studies have also addressed how symbiosis and water stresses interact to modify the function and expression of plant aquaporins.

Aquaporins are integral membrane proteins that function as gradient-driven water and / or solute channels present in plants. The effect of mycorrhiza interaction in aquaporin expression was described for several plant species, including the tree poplar (Marjanovic et al., 2005), though the microsymbiont was an ectomycorrhizal fungus. An aquaporin protein was recently cloned from an AMF, *Glomus intraradices*, by Aroca et al. (2009), but the interaction between the plant and fungi aquaporins remains unclear, as suggested by the authors, that the fungus aquaporin expression is due to plant. These results came from studies with culture plant species, opening a new view to investigate the expression profile of aquaporins from tree species and its rules at different ecosystems. An aquaporin (water channel) gene from an AM fungus (*Glomus intraradices*), which was named GintAQP1, was reported from experiments in different colonized host roots growing under several environmental conditions. It seems that gene expression is regulated in a compensatory way regarding host root aquaporin expression (Aroca et al., 2009). At the same time, from in vitro experiments, it was shown that a signaling communication between NaCl-treated mycelium and untreated mycelium took place in order to regulate gene expression of both GintAQP1 and host root aquaporins. The authors suggested that specific communication could be involved in the transport of water from osmotically favorable growing mycelium or host roots to salt-stressed tissues.

Despite their importance to plant productivity and sustainability of agricultural systems (Barea, 1991; Smith & Read, 1997), AMF are widely distributed through the most diverse forest ecosystems, as rain forest in southern Queensland-Australia (Gehring & Connell, 2006), Clintonia borealis roots from a boreal mixed forests in northwestern Québec (DeBellis & Widden 2006), hot-dry valley of Jinsha River, southwest China (Dandan and Zhiwei, 2007), forest with *Araucaria angustifolia* in Brazil (Moreira et al., 2007; Moreira et al. 2009; Patreze et al., 2009), in the Atlantic Forest in Southeastern Brazil (Zangaro et al., 2007, 2009), *Hepatica nobilis* Mill. site type spruce forest at central Estonia (Uibopuu et al., 2009), semi-evergreen tropical forest at southeast Mexico (Ramos-Zapata et al., 2011a), *Podocarpus cunninghamii* forests from New Zealand (Williams et al., 2011), young and old secondary forest in Western Brazilian Amazon (Stürmer & Siqueira, 2011), coastal dunes of Sisal, Mexico (Ramos-Zapata et al., 2011b) and floodplain islands as recently related at northeastern Italy (Harner et al., 2011). The observations of AMF occurrence in aquatic environment corroborate the hypothesis of coevolution with the first established plants (Berbee & Taylor, 1993).

Arbuscular mycorrhizas are abundant in herbaceous species as well as in tropical and temperate forest tree species (Harley & Harley, 1987; Newman et al., 1994). The temperate and boreal forest ecosystems are the best studied; however, AM are poorly investigated in these ecosystems since the most of tree species form ectomycorrhizas, a fungal polyphyletic group. In the Pinaceae family, for example, very common in temperate forests, a few species are able to form both ecto- and arbuscular mycorrhizas, e.g., *Pinus muricata, Pinus banksiana, Pinus strobus, Pinus contorta* and *Picea glauca x Picea engelmannii* (O'Dell et al., 1993; Horton et al., 1998; Wagg et al., 2008). Another temperate tree species, Douglas fir, showed over 200 morphologically distinct ectomycorrhizas in southern Oregon (Luoma et al., 1997). The authors reported that after disturbance such as fire burning or logging, roots of Douglas fir seedlings can also be colonized by arbuscular mycorrhizas.

On the other hand, the tropical forests are an outstanding biodiversity hotspot for vascular plants and consequently it is expected a high fungal diversity, including the AM which are symbiotic organisms of vascular plants. These fungi are important as they play a key role for nutrient cycling and nutrient retention in the humus layers. Then, tropical forests rich in plant species from diverse families are considered to be dominated by AM-forming trees; however, relatively few studies have focused at the molecular AM diversity in natural ecosystems and this knowledge is a major bottleneck in mycorrhizal ecology.

In tropical areas, most AMF propagules have shown seasonal fluctuations in abundance either as spores (Guadarrama & Alvarez-Sánchez, 1999, Silva-Júnior & Cardoso, 2006) or in colonized roots (Ramos-Zapata et al., 2011b). The approach which has been conducted is the evaluation of soil quality by direct counts of spores extracted from soil (Carvalho et al., 2003; Moreira et al., 2009), through assessment of percentage of colonized roots (Moreira et al., 2007, Patreze et al., 2009), and estimation of length and biomass of hyphae in the soil using hyphal ^{32}P-labelling (Pearson & Jakobsen, 1993). According to Rosendahl (2008), quantitative studies of arbuscular mycorrhizal fungal communities based on the presence of spore numbers are complicated as some species produce few spores on the mycelium, whereas species such as *G. intraradices, G. versiforme* or *G. fasciculatum* produce hundreds of spores on the same hypha. Another technique, the Most Probable Number (MPN) method is a microbiological approach that allows the detection of AMF species which do not produce spores (Troeh & Loynachan, 2003) by soil dilution. However, there are several experimental variables which may influence the final estimations. Several experiments have been carried out at greenhouse for evaluation of dependency and responsiveness to arbuscular mycorrhizal fungi in tree species, as cedar, *Cedrela fissilis* Vell. (Rocha et al., 2006), that occurs in different biomes from Brazil. Siqueira et al. (2010) published recently a book summarizing 30 years of research with mycorrhizal fungi in Brazil. The mycotrophic tree species was discussed in respect to potential use for restoration of degraded land (Soares & Carneiro, 2010) in tropical and arid ecosystems. The use of mycotrophic species in agroforestry systems was also discussed and several examples were discussed, as the studies for peach palm (*Bactris gasipaes* Kunth) and cupuaçu (*Theobroma grandiflorum* (Willd ex Spring) K. Schum) at the central part of the Amazon region in Brazil (Silva-Júnior & Cardoso, 2006).

We propose in this chapter to discuss the advances in molecular diversity of Glomeromycota in forest ecosystems focusing on some aspects related to DNA target regions for sequencing tools, on different approaches on molecular diversity applied to fungal researches and the presence of Glomeromycota in natural and impacted forests. Finally, we will address the challenges to the development of new areas, as genomic and metagenomic applied to mycorrhizal studies.

2. New approaches from molecular techniques

Studies in planta have shown the inability to obtain axenic cultures and the difficulties associated with identifying AMF, made more difficult to establish in the past, advanced studies on their ecology, genetics, and evolution. In the past decade, considerable effort has been expended to understand the keystone ecological position of AM symbioses, most studies have been limited in scope to recording organism occurrences and identities, as determined from morphological characters and ribosomal sequence markers for characterization of AMF, leading to important advances in our understanding of the phylogeny (Schüssler et al., 2001; Schwarzott et al., 2001), ecology (Helgason et al., 1998; Helgason et al., 2002; Husband et al., 2002a, b; Kowalchuk et al., 2002), genetics (Gianinazzi-Pearson et al., 2001; Harrison, 1999), and evolution (Gandolfi et al., 2003; Sanders, 2002) of this group of obligatory symbiotic fungi. rRNA genes have become the most widely use targets for detection of AMF in environmental samples (Clapp et al., 2002). Several PCR-based strategies targeting rRNA genes have more recently been developed to detect AMF in DNA extracted from roots, soil, or spores (van Tuinen et al., 1998; Helgason et al., 1998; Kjoller & Rosendahl, 2000; Kowalchuk et al., 2002; de Souza et al., 2004). Gamper et al. (2010) proposed a shift toward plant and fungal protein-encoding genes as more immediate indicators of mycorrhizal contributions to ecological processes. A number of candidate target genes, involved in the uptake of phosphorus and nitrogen, carbon cycling, and overall metabolic activity were proposed, and advantages and disadvantages of future protein-encoding gene marker and current (phylo-) taxonomic approaches are offered as new strategy for studying the impact of AM fungi on plant growth and ecosystem functioning.

Molecular approaches to community ecology may minimize data variation in the morphological characters that hamper traditional taxonomy and have revealed a considerable unknown AM diversity from colonized roots (Rosendahl, 2008) and soils under different land uses. In forest ecosystems, different groups of fungi, bacteria, algae, and microfauna communities living within the first soil layers can be altered by several factors. The relationship between diversity of fungal communities and resource available and the relation of fungal communities to the greater plant diversity remains under discussion. Using the Ribosomal Intergenic Spacer Analysis (RISA), Waldrop et al. (2006) found no significant effect of plant diversity on the number of fungal ITS bands. However, many other factors unrelated to plants, but inherent to soil (climate, parent material, slope), may influence fungal diversity and they are not easily controlled (Waldrop et al., 2006). The opposite also is possible, i.e. the existence of effects of AM on plants composition and development of plant communities, despite your action to the nutritional status of individual plants (Grime et al., 1987; van der Heijden et al., 1998).

The first steps to better understand the plant host-arbuscular mycorrhizal fungus (AMF) interaction in forest ecosystems could be the deeper studies on plant growth response to different natural soil inocula and upgrading of knowledge about the AM species composition from each community. Williams et al. (2011) observed that pre-inoculation of tree seedlings of *Podocarpus cunninghamii* propagated in glasshouse from cuttings with forest AMF-inoculums collected from a remnant *P. cunninghamii* forest could improve restoration success in comparison to the ex-agricultural AMF community used as inoculums. This last community was less mutualistic than the forest AMF community. These results have potential implications for forest restoration, predicting for example the effect of future forest management on understory forest vegetation. The molecular tools might complement such data helping the identification of mycorrhizal species in different forest communities. An accurate assessment of species richness and community composition is crucial to understanding the role of AMF in ecosystem functioning.

2.1 DNA target regions for sequencing tools

Molecular techniques were developed primarily for the identification of ectomycorrhizas (Gardes et al., 1991; Gardes & Bruns, 1993), and later for the analysis of arbuscular mycorrhizal fungi (AMF) (Lanfranco et al., 1999; Schüssler et al., 2001). Ectomycorrhizal fungi are a large diverse group of an estimated 5000–6000 different species belonging to the Basidiomycota and Ascomycota (Molina et al., 1992), which are very common in forest ecosystems, mainly in temperate and boreal forest ecosystems, where most tree species form ectomycorrhizas (Ducic et al., 2009).

All ribosomal genes (rRNA) are conserved in eukaryotes genomes and they are present in tandem repetition. The regions 18S, 5.8S and 28S are the most conserved, which allows the primer design to amplify the variable and informative sequences (internal transcribed spacer -ITS and intergenic spacer rRNA –IGS).

The fungal internal transcribed spacer (ITS) region of genomic DNA was characterized from single AMF spores by restriction fragment length polymorphism analysis (PCR-RFLP) (Sanders et al., 1995). Then, the ITS region was used to detect AMF in different roots systems (van Tuinen et al., 1998; Colozzi-Filho & Cardoso, 2000; Redecker, 2000) and in the field (Renker et al., 2003, Mergulhão et al., 2008). Using specific PCR primers to identify AMF within colonized roots of *Plantago media* and *Sorghum bicolor*, Redecker (2000) defined five groups of AMF. Later on, the same groups were detected by Shepherd et al. (2007) in roots of twelve tree legumes and non-legume trees, but these primers did not discriminate the AMF species. A set of primers amplifying a SSU-ITS-LSU fragment was developed (Kruger et al., 2009) allowing phylogenetic analyses with species level resolution. Such primers are useful to monitor entire AMF field communities, but they present a drawback related to their size of 1500 bp. Candidate regions to be DNA barcoding of arbuscular mycorrhizal fungi were analysed (Stockinger et al., 2010), but there was intraspecific variation heterogeneous and high in some groups.

Glomeromycota has a distribution of ITS fragment lengths concentrated between 550 and 650 bp, found in 96.4% of in silico analyzed sequences by Patreze et al. (2009). These authors rescued eight Glomeromycota genera and 31 species from 422 ITS sequences. The sub-regions ITS1 and ITS2 show high evolution rate and they are typically specific-species (Bruns & Shefferson, 2004). Furthermore, the great number of ITS copies per cell (more than 250) characterize this region as good target to sequencing where the DNA initial quantity is low, as environmental samples (Nilsson et al., 2009a). Due some taxonomic discrepancies between ITS1 and ITS2 analyzed separately and also in relation to the full ITS region (Nilsson et al., 2009b), it is suggested starting the study using the ITS2 region because it is as variable and long as ITS1, but ITS2 (White et al., 1990) has a major number of access at INSD (International Nucleotide Sequence Databases; Benson et al., 2008) to perform comparisons. Although the ITS sequencing allows species identification, the number of samples required for environmental studies can be unviable. The region ITS1-5.8-ITS2 was used to assess the genetic diversity of geographical isolates of *Glomus mosseae* (Avio et al., 2009).

The target regions Large Subunit (LSU) rDNA and the Small Subunit (SSU) rDNA have been useful at AMF's detection, whereas most studies are based on this last one. The resolution of these genes is different, and a direct comparison of phylogenetically defined taxa is not possible. Van Tuinen et al. (1998) detected four AMF species using one region from LSU rDNA in stained mycorrhizal root fragments by nested PCR. Species and in some cases isolates, also were separated based on polymorphism found in the gene coding for the large ribosomal subunit (LSU) by Single Stranded Conformation Polymorphism (SSCP) method (Kjøller and Søren Rosendahl, 2000). In this later method, nucleotide differences between homologous sequence strands are detected by electrophoresis of single-stranded DNA under non-denaturing conditions (Orita et al., 1989). The question was if it was possible to apply to field roots with unknown arbuscular mycorrhizal symbionts. Some years after, Rosendahl & Stukenbrock (2004) used with success the LSU rDNA sequences to analyze the community structure in coastal grassland in Denmark. The LSU provides a better resolution, but several primers are necessary for amplifying all genera of Glomeromycota. Lee et al. (2008) developed new primers using the small subunit rRNA gene (SSU) as target, providing another alternative to detect AMF directly from field roots. The sequencing of DNA target regions can reveal high variability of taxon richness and composition between particular ecosystems. Öpik et al. (2006) surveyed 26 publications that report on the occurrence of natural root-colonizing AM fungi identified using rDNA region (ITS, SSU and/or LSU), of which nine reports were in forest ecosystems. The number of AM fungal taxa per host plant species in tropical forests was 18.2 and temperate forests were 5.6. The Table 1 summarizes the results of surveys about AMF detection using rDNA regions in forest ecosystems after 2006. Data from two reports were obtained using second-generation sequencing technologies where taxa at very low abundances may be recorded.

Forest ecosystem	Plant species	N° of root samples screened	N° of clones screened	N° of AMF sequenced	OTU	Diversity index	Marker region	Primers used	Reference
seminatural woodland, North Yorkshire	*Hyacinthoides non-scripta*	33	141	62	**	**	SSU	NS31/A M1	Helgason et al. 1999
	Cajanus cajan	5			48	2.67			
	Heteropogon contortus	5			24	1.88			
boreonemoral forest, Central Estonia	10 species	458	*	158 358	47	9.96 to 38.32	SSU	NS31/A M1	Opik et al. 2009
gypsum area, Southern Spain	*G. struthium* L.	24	3072	1443	19	1.13	SSU	NS31 and AM1+AM2+AM3	Alguacil et. al 2009
	Teucrium libanitis Schreber	24						Nested PCR (NS31 / NS41 and ARCH1311/NS8)	
	Ononis tridentata L.	24							
	Helianthemum squamatum (L.) Dum.Cours	24							
hot and arid valley, Southwest China (undisturbed land)	*Bothriochloa pertusa*	5	1168	241	25	2.38	LSU	Nested PCR (LR1/FLR2 and 28G1/28G2)	Li et al. 2010
northern hardwood forests, Michigan, USA	maple (*Acer* spp.) roots	144	2160	38	27	1.94	18S	AM1/NS31	Van Diepen et al. 2011
mosaic of grassland, wood and heath, UK	soils (area of 7 m²)	66 soil cores	*	108 245	70	2.45	SSU	Nested PCR (NS31/A M1 and WANDA/AM1)	Dumbrel et al. 2011

Table 1. Overview of arbuscular mycorrhizal (AM) fungal community surveys from forest ecosystems. Data from sampling and sample screening were included. The asterisk means that the manuscript applied the pyrosequencing approach.

The fungal communities analyses by sequencing is based on PCR amplification using specific primers for the taxa in study, followed of cloning of fragments which represent the species richness. This kind of approach generates a library of clones, which many times are high (above one hundred). In order to minimize costs, the clones obtained can be select by restriction fragment length polymorphism (RFLP), grouped according a restriction standard revealed using restriction enzymes. Helgason et al. (1998) made use of this technique for the first time to evaluate the AMF diversity changes comparing agriculture soils and forest adjacent. They suggested that the low taxonomic diversity of arbuscular mycorrhizal fungi in arable fields indicates that their functional contribution may be less there than in woodland.

To date, almost all information on sequence differences in this interesting fungal group comes from ribosomal genes. Other coding regions of the genome were investigated as the variability of β-tubulin and H+-ATPase genes in the AMF *Glomus intraradices* (Corradi et al., 2004). For this purpose, the authors used degenerate primers in order to sequence the most gene variants possible including any that might have originated from other fungal and eukaryotic groups. Following this idea, it is important to check the sequences available on databases of additional fungal groups to improve the consistence of phylogenetic analysis for arbuscular mycorrhizal fungi, mainly when the objective is evaluate the variability in other than ribosomal genes.

2.2 Molecular methods for fungal diversity and applications

Many methods allow the elucidation of microbial structure has been intensively applied to bacterial and fungal communities, as PCR-Restriction Fragment Length Polymorphism (PCR-RFLP), Denaturing Gradient Gel Electrophoresis (DGGE), Temperature Gradient Gel Electrophoresis (TGGE), Terminal-Restriction Fragment Length Polymorphism (T-RFLP), or Oligonucleotide Fingerprinting of rRNA Genes (OFRG). Here, we select studies which were developed from these methods in Glomeromycetes communities which were applied or have a potential for assessment in natural ecosystems, as forests. All possibilities are described in Figure 1, in an adaptation from Theron & Cloete (2000). The actual function of AMF symbiosis in nature should be considered at the community level of both the AMF and host plants, but we are focusing at fungal partner. All primers sequences and regions of ribosomal RNA genes used in such molecular approaches are shown at Figure 2.

2.2.1 PCR-RESTRICTION Fragment Length Polymorphism (PCR-RFLP) and Single Stranded Conformation Polymorphism (SSCP)

The technique PCR-RFLP was employed with success (Sanders et al., 1995) to distinguish AMF species from DNA isolated of spores; however, when applied to field samples, this technique can generate polymorphism in not target organisms. Avio et al. (2009) were able to discriminate *Glomus mosseae* isolates from G. coronatum, *G. intraradices* and G. viscosum by using of a single enzyme (HinfI) with ITS-RFLP profiles. For field samples, this technique remains insufficiently tested, Mergulhão et al. (2008) detected AMF species in an impacted semiarid soil using the ITS1-5.8S-ITS2 region and Börstler et al. (2010) analyzed for the first time the intraspecific genetic structure of an AMF directly from colonized roots in the field comparing between agricultural and semi-natural sites. To our knowledge, there are not studies using solely PCR-RFLP to characterize AMF communities in forest ecosystems. The work from van Diepen et al. (2011) used the PCR-RFLP to select clones representatives of each type to be re-amplified and sequenced.

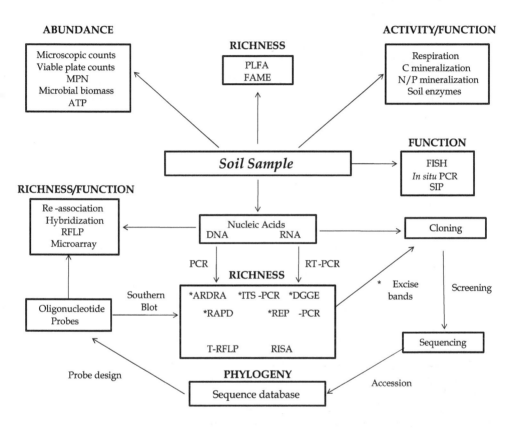

Fig. 1. Commonly used molecular approaches in microbial ecology. PLFA - Phospholipid Fatty Acids; FAME – Fatty Acid Methyl Ester; FISH – Fluorescence in Situ Hybridization; SIP - Stable Isotope Probing; RFLP - Restriction Fragment Length Polymorphism; PCR - Polymerase Chain Reaction; RT-PCR – Real Time Polymerase Chain Reaction; ARDRA – Amplified Ribosomal DNA Restriction Analysis; ITS-PCR – Internal Transcribed Spacer Polymerase Chain Reaction; DGGE – Denaturing Gradient Gel Electrophoresis; RAPD – Random Amplified Polymorphic DNA; T-RFLP – Terminal Restriction Fragment Length Polymorphism; REP-PCR – Repetitive Element Palindromic Polymerase Chain Reaction; SIP – Stable Isotope Probing.

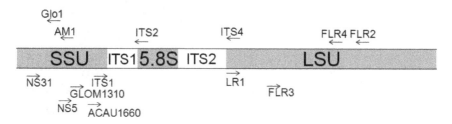

Primer	Sequence (5'-3')	Reference
ITS1	TCCGTAGGTGAACCTGCGG	White et al. 1990
ITS2	GCTGCGTTCTTCATCGATGC	White et al. 1990
ITS4	TCCTCCGCTTATTGATATGC	White et al. 1990
NS5	AACTTAAAGGAATTGACGGAAG	White et al. 1990
NS31	TTGGAGGGCAAGTCTGGTGCC	Simon et al. 1992
LR1	GCA TAT CAA TAA GCG GAG GA	van Tuinen et al. 1998
AM1	GTTTCCCGTAAGGCGCCGAA	Helgason et al. 1998
FLR2	GTC GTT TAA AGC CAT TAC GTC	Trouvelot, et al. 1999
ACAU1660	TGAGACTCTCTCGGATCGG	Redecker et al. 2000
GLOM1310	AGCTAGGYCTAACATTGTTA	Redecker et al. 2000
FLR3	TTGAAAGGGAAACGATTGAAGT	Gollotte et al. 2004
FLR4	TACGTCAACATCCTTAACGAA	Gollotte et al. 2004
Glo1	GCCTGCTTTAAACACTCTA	Cornejo et al. 2004

Fig. 2. Diagram of the ribosomal DNA cluster of fungi containing the location and sequences of the primers cited in this chapter and used in molecular methods to fungal community studies. The ribosomal RNA (rRNA) gene is a tandem array of at least 50–100 copies in the haploid genome of all fungi. SSU=small ribosomal subunit; ITS1=intergenic transcribed spacer region 1; ITS2=intergenic transcribed spacer region 2; LSU=large ribosomal subunit. The arrows refer to the approximate annealing sites of primers, but the diagram is not to scale.

In order to detect AMF species, nested-PCR, based on sequence differences in the gene coding for the large ribosomal subunit can be coupled to a method known as SSCP (Single Stranded Conformation Polymorphism) and the differences among species are visualized in polyacrylamide gels under non-denaturing conditions. This method allowed Kjøller & Rosendahl (2000) detect four species of Glomus in root tissues of four culture plant species. Jansa et al. (2002) could detect different ITS sequences types within of the single-spore isolates of *Glomus intraradices* using the SSCP technique. A recent study assessed the AMF community at field, in arid gypsophilous plant communities in south-eastern Spain (Alguacil et al., 2009) by sequencing (Table 1), using SSCP to select clones. In such work, representatives of each SSCP pattern were chosen for sequencing while the remaining clones (almost two thousand) were classified by SSCP typing. These sequences showed high degree of similarity to sequences from taxa belonging to the phylum Glomeromycota. In a savanna area, Alguacil et al. (2010) related the AMF diversity in roots of *Centrosema macrocarpum* to soil parameters and sources of phosphorus. The authors amplified by PCR the AM fungal small-subunit (SSU) rRNA genes and selected clones by SSCP to sequencing and phylogenetic analyses. They identified nine fungal types: six belonged to the genus Glomus and three to Acaulospora. The single-stranded conformation polymorphism (SSCP) approach is a very sensitive and reproducible technique that has potential to be applied successfully in studies in order to analyze the sequence diversity of AM fungi within roots from forest ecosystems, not reported so far.

2.2.2 Terminal-Restriction Fragment Length Polymorphism (T-RFLP)

Another technique, derived from a combination of PCR, RFLP and electrophoresis of nucleic acids (Liu et al., 1997) nominated as Terminal-Restriction Fragment Length Polymorphism (T-RFLP) use oligonucleotides fluorescent-labeled that enables generation of fingerprinting data of microbial communities efficiently (Marsh, 1999). Figure 3 exemplifies the steps needed for the T-RFLP analysis. As PCR-RFLP, the selection of restriction enzymes is a fundamental step. It is important to use 2-4 enzymes in each study for obtaining of different amplicons (Tiedje et al., 1999). These authors recommend the enzymes HhaI, RsaI, and MspI, which have given the greatest resolution based on restriction analysis of the database as well as natural communities for soil microbial, but others may be of use under special circumstances, as specific AMF communities. The initial works using T-RFLP for fungal community analysis focused on temperate forests (Klamer et al., 2002) or ectomycorrizas (Zhou and Hogetsu, 2002; Dickie et al., 2002). Combining LSU rDNA sequencing and T-RFLP analysis, Mummey et al. (2009) investigated if the pre-inoculation may play a role in arbuscular mycorrhizal fungi (AMF) community assembly within the roots. Another application of T-RFLP for arbuscular mycorrhizal can be developed to measure the effect of soil inoculums representing different AM fungal communities on the growth of three plant species. Uibopuu et al. (2009) used Glomeromycota specific primers NS31 and AM1 labeled with fluorescent dyes to perform the method known as T-RFLP, comparing the inoculums from a young forest stand, an old forest stand and an arable field at growing of the three plant species and showed that the old and young forest resulted in similar root AMF communities whilst plants grown with AM fungi from arable field hosted a different AMF community from those grown with old forest inocula. However the AMF richness in plant roots was not related to the origin of AMF inoculums. Previous works using the T-RFLP technique were performed in AMF community colonizing roots from herbaceous (Wu et al. 2007), grass (Mummey et al., 2005, Mummey and Rillig, 2008; Hausmann and Hawkes, 2009), addressing studies of impact of various agricultural practices on AMF biodiversity (Lekberg et al., 2007; Verbruggen et al., 2010). The average number of AMF taxa reached in this work was 8.8 OTU and the authors stressed the importance of organic management in agro-ecosystems maintenance of mycorrizal fungi. Van de Voorde et al. (2010) compared the AMF communities from roots of one species of interest in both situations: bioassay plants and plants collected from the field. Although the species had not been a forest, wood or tree plant, this kind of study can be applicable to address similar questions for them.

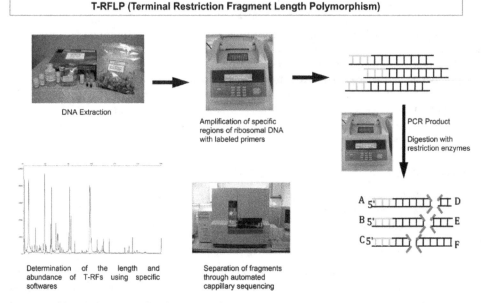

Fig. 3. A schematic design for the steps of a T-RFLP analysis on the soil microbial community structures or for determination of the presence of a target species using specific primer.

Once the belowground feedbacks may lead to changes in species diversity, the knowledge about AMF diversity and distribution both in soil and roots, as well as plant performance is very useful to sustainability of an ecosystem. Evaluating the AMF diversity in roots from seven different shrub species, Martínez-García (2011) suggested that the generate islands of fertility which differ in nutrient content and, therefore, support different AMF communities, increasing AMF diversity at the landscape level. These authors used the primers sequences LR1 and FLR2 for the amplification of the 5' end of LSU rDNA sequences in general fungi (Van Tuinen et al., 1998; Trouvelot et al., 1999) and in the second amplification, they used the AMF specific primers FLR3 and FLR4 (Gollotte et al., 2004). The same set of primers were used by Koch et al. (2011) to investigate the impacts of introduced plants and exotic AM fungi on local AM fungi at Canada.

2.2.3 Denaturing Gradient Gel Electrophoresis (DGGE), Temperature Gradient Gel Electrophoresis (TGGE)

Denaturing gradient gel electrophoresis (DGGE) and terminal-restriction fragment length polymorphism (T-RFLP) approaches also were applied to AMF studies. These techniques are based on extraction of community DNA followed by the PCR amplification of rRNA genes from the community DNA using universal, domain or group specific primers. The resulting products are separated based on relative helix stabilities in a denaturant (DGGE) or thermal (TGGE) gradient gel (Muyzer et al., 1993). Such techniques are very sensitive and have been used to detect single base differences. However, the gel system employed has low resolving power and there is no way of defining with accuracy the Tm of the helix.

Moreover, there is no comparative sequence database for AMF. In addition, reports have been published and stimulate to better focus on this issue. Kowalchuk et al. (2002) first coupled an AMF-specific PCR strategy targeting the 18S rRNA gene (Helgason et al., 1998) with denaturing gradient gel electrophoresis (DGGE) to detect AMF in roots of a grass species at two coastal sand dune locations in the Netherlands. In this report, the primer reverse NS31, described by Simon et al. (1992) had a GC clamp sequence, as described by Kowalchuk et al. (1997). The same gene (18S rRNA), though a different sub-region, was used by de Sousa et al. (2004) who applied the DGGE method to discriminate Gigaspora isolates and Gigasporaceae populations from environmental samples. Although the environmental samples were collected in a grassland field, a cattle farm in Brazil, that work brings new possibilities for studying the ecology of Gigaspora under field conditions, including forest ecosystems, without the need for trap cultures.

The DGGE from sequential amplification of 18S rDNA fragments by nested PCR using primer pairs AM1-NS31 and Glo1-NS31GC yielded a high-resolution band profile to soil samples from different ecosystems, including a eastern red cedar (Juniperus virginiana L.) forest ecosystems (Liang et al., 2008). The primer Glo1 was described by Cornejo et al. (2004). Although the primer pair AM1/NS31 is one of the most widely used group-specific primer pairs in studies of AMF communities, amplifying the three well-established families of the Glomeromycota (Glomaceae, Acaulosporaceae, and Gigasporaceae), this primer pair does not amplify 18S rDNA fragments from all known AMF or they can amplify some non-AMF sequences. Nevertheless, an advantage for the use of primer pair AM1/NS31 is the relatively large amounts of DNA sequence information derived from this primer pair available. A subsequent study (Zhang et al., 2010) was aimed the variable V3-V4 region of the 18S rDNA of AMF gene, by using nested PCR in three steps: (1) first round PCR, using primers GeoA2 and Geo11 (Schwarzott and Schüssler, 2001); (2) second round, using primers above-mentioned AM1/ NS31-GC; and (3) third round using NS31-GC/Glo1. In that study, the AMF community from rhizosphere of two shrubs species was investigated and the species richness ranged from 17 to 25 AMF species. Internal Transcribed Spacer (ITS) specific primers for Acaulosporaceae (ACAU1660/ITS2) and Glomaceae (GLOM1310/ITS2) (Redecker et al., 2000; White et al., 1990) have been used successfully in DGGE analysis on differentiating of composition of mycorrhizal communities in maize genotypes (Oliveira et al., 2009; Pagano et al., 2011).

Two other studies aiming at the molecular community analysis of AMF had as target the fungal small subunit (SSU) rRNA gene. Their objectives were related to role of AMF in plant tolerance to heavy metals stress (Long et al., 2010) and the interplay between soil properties and crop yield (Wu et al., 2011).

Few studies have applied molecular tools as DGGE analysis in forest species. The scarcity of works of this nature in forest species reinforces our goal of encouraging research in this area. Öpik et al. (2003) had surveyed the mycorrhizal status of plants grown in soils from a boreal forest by DGGE plus restriction analysis and sequencing. The region analyzed by them was the SSU region. Recently, using specific PCR conditions for Glomaceae family (nested system with NS5/ITS2 and GLOM1310/ITS2 primers) in DGGE system, Pagano et al. (2011) showed the applicability of this technique to understand the role of AMF in woody and shrub species from Caatinga, a dry deciduous forest at Brazil. The region studied had different agroforestry systems which were implanted in a degraded area in order to be an attractive alternative to conventional afforestation systems. An important conclusion from this work is the existence of functional diversity among AMF, supporting the theory that the AMF are considered as one of the factors that determine how plant species coexist. As observed by Pagano et al. (2011) for semi-arid soil, the analysis of AMF population of an experimental area may inform the state of land restoration depending how close they are from those of climax vegetation.

2.2.4 F-RISA and Automated-RISA (ARISA)

The Fungal Ribosomal Intergenic Spacer Analysis (F-RISA) method exploits the variability on the length of the nuclear ribosomal DNA (rDNA) region that contains the two internal transcribed spacers (ITS1 and ITS4) and the 5.8S rRNA gene (ITS1-5.8S-ITS2). Gleeson et al. (2005) characterized the fungal community structure on mineral surface using this region and Hong, Fomina and Gadd (2009) showed the applicability of this assay to examine the potential role of fungi as bioindicator of effects of organic and metal contamination in soil. It is possible to use the same approach to Glomeromycetes, but it is suggested to sequence some F-RISA fragments from AMF species known to establish standards. Thus, differential fragments when experimental communities are compared can be excised from gel, purified and sequenced in order to detect core AMF species in each environment or ecosystem.

To improve the resolution of this technique was developed an automated variation (ARISA) by Fisher & Triplett (1999) for characterization of bacterial communities. This PCR-based technique is based on the use of a fluorescent primer in the amplification of microbial ribosomal intergenic spacers, using DNA extracted from environmental samples as a template. ARISA was first used to fungal soil communities by using a pair of primers that targeted the 3′end of the 18S rDNA sequence and the 5′end of the 25S rDNA sequence (Ranjard et al., 2001). These authors examined the fungal database for the size of the ITS1-5.8S-ITS2 region in fungi, totalizing 104 genera and 251 species. However, the Glomeromycetes were not recognized as a phylum that date. Patreze et al. (2009) repeated the same in silico analysis including data updated to January 2008. The authors followed the classification of Hibbett et al. (2007), which consider the Glomeromycota as a phylum. Representatives of this phylum have a distribution of ITS fragment lengths concentrated between 550 and 650 bp (Patreze et al., 2009). The authors concluded that a clear distinction among the fungi kingdom is not possible considering the ITS sequence length. However, the method RISA was useful to characterize soil fungal communities from three forest ecosystems from Brazil: a native forest of *Araucaria angustifolia* and two replanted forest.

2.3 Microsatellites

Genetic diversity of arbuscular mycorrhizal fungi also can be investigated from the viewpoint of the population or individual, aside from the community level. Multilocus genotyping of AMF using microsatellites have been useful as marker suitability for population genetics. Microsatellites or Simple Sequence Repeats (SSRs) are regions with at least five identical repeats of two, three or four nucleotides, or a stretch of at least 10 identical single nucleotides. The length polymorphisms at microsatellite regions are caused by changes in the numbers of repeat lengths, which are repeated up to about 100 times (Tautz, 1989). This marker was used to explore the AMF diversity, simultaneously published by Croll et al. (2008a) and Mathimaran et al. (2008a) for the *Glomus intraradices* species. Previous works of distinct nature had reveled genetic variation within AMF species (Vandenkoornhuyse & Leyval, 1998), which affect plant growth and nutrition (Koch et al., 2006). Previously, the possibility of using a tandem repeated DNA sequence as a diagnostic probe for detection in colonized roots was demonstrated from the arbuscular mycorrhizal fungus *Scutellospora castanea* (Zézé et al., 1996). Then Zézé et al. (1997) employed the M13 minisatellite-primed PCR technique to explore the intersporal genetic variation of *Gigaspora margarita*. In the same year, the microsatellites were used as target to detect mycorrizal fungi (ecto and endo-mycorrhiza), including AMFs, however the isolates of *Glomus mosseae* could not be separated by microsatellites analysed (Longato & Bonfante, 1997). In addition, Douhan and Rizzo (2003) had developed a technique to isolate and detect microsatellite loci in AM fungi from single spores of *Glomus etunicatum* and *Gigaspora gigantea*. The authors were not certain that the microsatellite motifs found by them were from the target organism due the possible contaminants. A fingerprinting technique widely used in studies of closely organisms (Lim et al., 2004) known as Inter-Simple-Sequence Repeat (ISSR-PCR) allowed

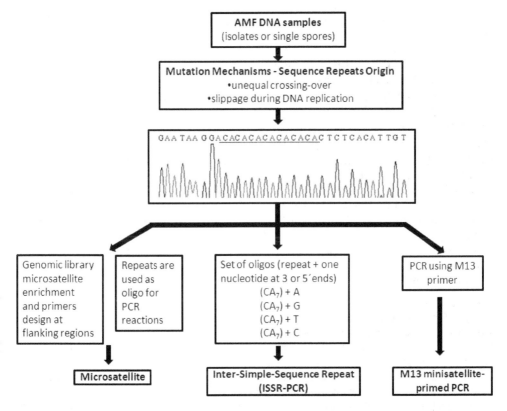

Fig. 4. Microsatellites regions origin and different approaches to explore the variation at length repeats: Microsatellite or Simple Sequence Repeats; Inter-Simple-Sequence Repeat and M13 minisatellite-primed PCR. These techniques were applied to arbuscular mycorrhizal fungi studies.

the characterization of genetic diversity of *Glomus mosseae* isolates (Avio et al., 2009). The authors amplified the minisatellite and microsatellite regions from fungal extracted DNA using the primers M13, $(GTG)_5$ and $(GACA)_4$. The same strategy above-described has potential application to several AMF species. In our opinion, this study can be a precursor of a field research related to the functional diversity of arbuscular mycorrhizal fungi, which remains poorly explored.

Taking into account the AM fungal populations, the molecular techniques have also shown that natural populations exhibit unexpectedly high genetic diversity, despite the assumption that diversity in these seemingly asexual fungi should be low. Thus, microsatellites are an interesting alternative as markers since the ribosomal gene sequences in AM fungi present high diversity, which might cause problems in their use in field studies.

3. Conclusion

Advances at understanding of genome structure of AMF were done in specifics and common species, as *G. intraradices*, which can easily be cultured in large amounts. Analyzing the regions of both LSU rDNA (*Glomus intraradices*) and POL1-like (PLS) sequences (*Glomus etunicatum*), Boon et al. (2010) confirmed high intra-isolate genetic polymorphism at the genome level. Their showed that genetic variation persists at the transcript level and suggested that in AMF, multiple nuclear genomes contribute to a single phenotype. Thus, it is supposed AMF connect plants together by a hyphal network, and that these different genomes may potentially move around in this network. In addition, genomic changes do not only appear among highly divergent lineages but can also occur among highly related species and individuals from the same population (Corradi et al., 2006).

The reassociation kinetics on *G. intraradices* experiments conducted by Hijri and Sanders (2004) revealed that 1.59% from haploid genome size is repetitive DNA, the category that includes the microsatellites regions. However the repetitive regions had low frequency, these authors had suggested that the very small genome size of *G. intraradices* makes it an excellent candidate for a genome sequencing project, beyond this species to be one of the most commonly studied AM fungi which colonizes host plants rapidly. The production of a completely annotated and assembled *G. intraradices* genome was initiated in 2004, having been shown especially arduous challenge. Martin et al. (2008) summarized the main difficulties found to complete this project and presented a nice historical perspective about the advances and approaches used to sequencing.

The recently developed massively parallel ('454') pyrosequencing enables metagenomic and metagenetic analyses in a manner that increase the capacity of traditional Sanger sequencing-based approaches by several orders of magnitude (Tedersoo et al., 2010). Pyrosequencing of fungi in diverse environments, such as soil or roots, elevates the number of recovered taxa several fold (Table 1). Similarly to study of Tedersoo et al. (2010) which compares a ectomycorrhizal fungi community of a tropical rainforest ecosystem by pyrosequencing and Sanger's sequencing, such analysis need to be performed to arbuscular mycorrhizal fungi in order to improve technical biases and to interpreted the data accordingly. Recently, the suitability of species abundance models in arbuscular mycorrhizal fungi were addressed (Unterseher et al., 2011) using output data from a boreonemoral forest in Estonia (Öpik et al., 2009), described in the Table 1. The authors proposed the use of lognormal species abundance distributions (SAD) as a working hypothesis to elucidate MOTU richness and biodiversity of AMF communities with low to medium sampling coverage. Such analyses are recommended to new studies in AMF communities emerging from pyrosequencing.

Studies on the evolutionary ecology of the AMF are on demand in order to approach measuring selection and host specificity as variations in AMF phenotypes were observed more recently with the development of molecular techniques. New approaches based on protein-encoding genes are expected to open opportunities to advance the mechanistic understanding of ecological roles of mycorrhizas in natural and managed forest ecosystems as well. And the idea that direct selection on AM fungal traits related to their survival and performance in the environment independent of the host is being reviewed as extraradical mycelium can be shown to be responsible for a significant part of the diversity of the AM fungi. As proposed by Helgason & Fitter (2009), the fungal response to the abiotic environment is that it would be expected for there to be substantial uncharacterized diversity in the Phylum Glomeromycota, since its members are globally distributed but poorly dispersed, and soil conditions vary greatly in time and space.

There have been significant advances in the plant-microbe interaction studies. As example, laser microdissection (LMD), a method which has been used widely by human and animal biologists to study gene expression in specific cell types and to elucidate the associated molecular events, has been adapted to plant tissues (Day et al. 2005; 2006) and applied successfully to study root-mycorrhizal fungus interactions for the identification of differentially expressed transcripts from LMD-derived RNA for the development of the arbuscule–cortical cell interface (Gomez & Harrison, 2008) or identify transcripts of different phosphate transporters in the same arbusculated cell population provides (Balestrini et al., 2007). The special advantage of LMD for arbuscular mycorrhiza is the isolating of cortical cells containing the fungus from the rest of the root cells. One objective is to rescue the RNA to perform transcript profiles analysis. LMD opens a new scenario for the understanding of the molecular basis of the AM symbiosis. Although the preparation protocols needs to be optimized for each tissue type and plant species, LMD can be adapted to detect and quantify mycorrhizal fungus in forest roots.

4. Acknowledgment

The authors thank Conselho Científico e Tecnológico (CNPq), Fundação de Apoio à Pesquisa do Estado de São Paulo (FAPESP) and Fundação de Apoio à Pesquisa do Estado do Rio de Janeiro (FAPERJ) for financial support and fellowships.

5. Glossary

AMF = Arbuscular Mycorrhizal fungi
ARISA = Automated Variation
DGGE = Denaturing Gradient Gel Eletrophoresis
F-RISA = Fungal Ribosomal Intergenic Spacer Analysis
ITS = Internal Transcribed Spacer
LMD = Laser Microdissection
LSU = Large Subunit ribosomal of rRNA
OFRG = Oligonucleotide Fingerprinting of rRNA Genes

PCR = Polymerase Chain Reaction
RFLP = Restriction Fragment Length Polymorphism
RISA = Ribosomal Intergenic Spacer Analysis
rRNA = Ribosomal RNA
SSCP = Single Stranded Conformation Polymorphism
SSRS – Microsatellites or Single Sequence Repeats
SSU = Small Subunit ribosomal of rRNA
TGGE = Temperature Gradient Gel Eletrophoresis
T-RFLP = Terminal Restriction Fragment Length Polymorphism

6. References

Alguacil M. del M., Lozano Z., Campoy M.J., Roldán A. (2010) Phosphorus fertilisation management modifies the biodiversity of AM fungi in a tropical savanna forage system. *Soil Biology and Biochemistry,* Vol.42, pp.1114-1122. ISSN: 0038-0717

Alguacil M. M., Roldán A. Torres M. P. (2009) Assessing the diversity of AM fungi in arid gypsophilous plant communities. *Environmental Microbiology,* Vol.1, No.10, pp.2649–2659. ISSN: 1462-2912

Aroca R., Bago A., Sutka M., Paz J. A., Cano C., Amodeo G., Ruiz-Lozano J. M. (2009) Expression analysis of the first arbuscular mycorrhizal fungi aquaporin described reveals concerted gene expression between salt-stressed and nonstressed mycelium. *Molecular Plant-Microbe Interactions,* Vol.22, No.9, pp.1169–1178. *ISSN:* 0894-0282

Augé R. M. (2001) Water relations, drought and vesicular-arbuscular mycorrhizal symbiosis. *Mycorrhiza,* Vol.11, pp.3-42. *ISSN:* 1432-1890

Avio L., Cristani C., Strani P. Giovannetti (2009) Genetic and phenotypic diversity of geographically different isolates of *Glomus mosseae. Can. J. Microbiol,* Vol.55, pp.242-253. *ISSN:* 0008-4166

Azcón-Aguilar C., Cantos M., Troncoso A., Barea J. M. (1997) Beneficial effect of arbuscular mycorrhizas on acclimatization of micropropagated cassava plantlets. *Scientia Horticulturae,* Vol.72, pp.63-71. *ISSN:* 0304-4238

Balestrini R., Gómez-Ariza J., Lanfranco L., Bonfante P. (2007). Laser microdissection reveals that transcripts for five plant ando ne fungal phophate transporte genes are contemporaneously presente in arbusculated cells. *Molecular and Plant Microbe Interactions,* Vol.20, pp.1055-1062. ISSN: 0894-0282

Berbee M. L.,Taylor J. W. (1993) Dating the evolutionary radiations of the true fungi. Canadian Journal of Botany-Revue Canadienne de Botanique, Vol.71, No.8, pp. 1114-1127. *ISSN:* 0008-4026

Boon E., Zimmerman E., Lang B. F., Hijri M. (2010). Intra-isolate genome variation in arbuscular mycorrhizal fungi persists in the transcriptome. *J. Evol. Biol.,* Vol.23, pp.1519–1527. *ISSN:* 1420-9101

Börstler B, Thiéry O, Sýkorová Z, Berner A, Redecker D, 2010. Diversity of mitochondrial large subunit rDNA haplotypes of *Glomus intraradices* in two agricultural field experiments and two semi-natural grasslands. *Molecular Ecology,* Vol.19, pp.1497-1511. *ISSN:* 0962-1083

Bruns T. D., Shefferson R. P. (2004) Evolutionary studies of ectomycorrhizal fungi: recent advances and future directions. Canadian Journal of Botany-Revue Canadienne de Botanique, Vol.82, No.8, pp.1122-1132. *ISSN*: 0008-4026

Carvalho L. M., Correia P. M., Ryel R. J., Martins-Loução M. A. (2003). Spatial variability of arbuscular mycorrhizal fungal spores in two natural plant communities. *Plant and Soil*, Vol.251, pp.227–236. *ISSN*: 1573-5036

Clapp J.P., Helgason T., Daniell T.J., Young J.P.W. (2002). Genetic studies of the structure and diversity of arbuscular mycorrhizal fungal communities. *In*: M.G.A. van der Heijden and I.R. Sanders (ed.), Ecological Studies, vol. 157: Mycorrhizal Ecology. Springer Verlag, Berlin, German

Theron J, Cloete T.E. (2000). Molecular techniques for determining microbial diversity and community structure in natural environments. Crit. Rev. Microbiol.Vol. 26(1): pp. 37-57. ISSN-1040-841X

Colozzi Filho A., Cardoso E.J.B.N. (2000) Detecção de fungos micorrízicos arbusculares em raízes de cafeeiro e de crotalária cultivada na entrelinha. *Pesq. agropec. Bras.,*. Vol.35, No.10, pp.2033-2042. *ISSN* 1678-3921

Cornejo P., Azcon-Aguilar C., Barea J.M., Ferrol N. (2004) Temporal temperature gradient gel electrophoresis (TTGE) as a tool for the characterization of arbuscular mycorrhizal fungi. FEMS *Microbiology Letters*, Vol.241, pp.265–270. *ISSN*: 0378-1097

Corradi N., Kuhn G., Sanders I.R. (2004) Monophyly of b-tubulin and Hþ-ATPase gene variants in *Glomus intraradices*: consequences for molecular evolutionary studies of AM fungal genes. *Fungal Genetics and Biology*, Vol. 41, pp. 262–273. *ISSN*: 1087-1845

Corradi N., Sanders I. R. (2006) Evolution of the P-type II ATPase gene family in the fungi and presence of structural genomic changes among isolates of *Glomus intraradices*. *BMC Evolutionary Biology*, Vol. 6, No.1, pp. 21. *ISSN*: 1471-2148

Croll D., Wille L., Gamper H.A., Mathimaran N., Lammers P.J., Corradi N., Sanders I.R. (2008) Genetic diversity and host plant preferences revealed by simple sequence repeat and mitochondrial markers in a population of the arbuscular mycorrhizal fungus *Glomus intraradices*. *New Phytologist*, Vol.178, pp.672–687. *ISSN*: 1469-8137

Dandan Z., Zhiwei Z. (2007) Biodiversity of arbuscular mycorrhizal fungi in the hot-dry valley of the Jinsha River, southwest China. *Applied Soil Ecology*, Vol.37, pp.118-128. *ISSN*: 0929-1393

Day R.C., Grossniklaus U, Macknight R.C. (2005). Be more specific! Laser-assisted microdisssection of plant cells. *Trends Plant Science*, Vol.10, pp.397-405. *ISSN*: 1360-1385

Day R.C., McNoe L.A., Macknight R.C. (2006). Transcript analysis of laser microdissected plant cells. Technical focus, *Physiol. Plantarum*, Vol.129, pp.267-282. ISSN: 0031-9317

de Souza F.A., Kowalchuk G.A., Leeflang P., van Veen J.A., Smit E. (2004) PCR-Denaturing Gradient Gel Electrophoresis Profiling of Inter- and Intraspecies 18S rRNA Gene Sequence Heterogeneity Is an Accurate and Sensitive Method To Assess Species Diversity of Arbuscular Mycorrhizal Fungi of the Genus *Gigaspora*. *Applied and Environmental Microbiology*, pp.1413–1424. *ISSN*: 0099-2240.

DeBellis T., Widden P. Diversity of the small subunit ribosomal RNA gene of the arbuscular mycorrhizal fungi colonizing *Clintonia borealis* froma mixed-wood boreal forest. *FEMS Microbiol. Ecol.* Vol.58, pp.225–235. *ISSN*: 0168-6496

Dickie I. A., Xu B., Koide R. T. (2002) Vertical niche differentiation of ectomycorrhizal hyphae in soil as shown by T-RFLP analysis. *New Phytologist*, Vol.156, pp.527–535. *ISSN*: 1469-8137

Douhan G. W., Rizzo D. M. (2003) Amplified Fragment Length Microsatellites (AFLM) might be used to develop microsatellite markers in organisms with limited amounts of DNA applied to Arbuscular Mycorrhizal (AM) fungi. *Mycologia*, Vol.95, No.2, pp.368–373. *ISSN*: 0027-5514.

Ducic T., Berthold D., Langenfeld-Heyser R., Beese F., Polle A. (2009) Mycorrhizal communities in relation to biomass production and nutrient use efficiency in two varieties of Douglas fir (*Pseudotsuga menziesii* var. *menziesii* and var. *glauca*) in different forest soils *Soil Biology and Biochemistry*, Vol.41, pp.742–753. *ISSN*: 0038-0717

Fisher M. M., E. W. Triplett. (1999) Automated approach for ribosomal intergenic spacer analysis of microbial diversity and its application to freshwater bacterial communities. *Appl. Environ. Microbiol.*, Vol.65, pp.4630–4636. *ISSN*: 0099-2240

Gamper H.A., van der Heijden M.G., Kowalchuk G.A. (2010). Molecular trait indicators: moving beyond phylogeny in arbuscular mycorrhizal ecology. *New Phytologist*, Vol.185, No.1, pp.67–82. *ISSN*: 1469-8137

Gandolfi A., Sander I.R., Rossi V., Menozzi P. (2003) Evidence of recombination in putative ancient asexuals. *Mol. Biol. Evol.* Vol.20, pp.754–761.*ISSN*:0737-4038

Gardes M., Bruns T. D. (1993) ITS primers with enhanced specificity for basidiomycetes – application to the identification of mycorrhizae and rusts. *Molecular Ecology*, Vol.2, No.2, pp.113–118. *ISSN*: 0962-1083

Gardes M., White T.J., Fortin J.A., Bruns T.D., Taylor J.W. (1991) Identification of indigenous and introduced symbiotic fungi in ectomycorrhizae by amplification of nuclear and mitochondrial ribosomal DNA. Canadian Journal of Botany-Revue Canadienne de Botanique, Vol.69, No.1, pp.180–190. *ISSN*: 0008-4026

Gehring C. A., Connell, J. H. (2006) Arbuscular mycorrhizal fungi in the tree seedlings of two Australian rain forests: occurrence, colonization and relationships with plant performance *Mycorrhiza*, Vol.16, pp.89–98. *ISSN*: 1432-1890.

Gianinazzi-Pearson V., van Tuinen D., Dumas-Gaudot E., Dulieu H. (2001). Exploring the genome of glomalean fungi, p. 3-17. *In*: B. Hock (ed.), Fungal Associations, Vol. IX. The Mycota. Springer-Verlag, Berlin, Germany.

Gleeson D. B., Clipson N., Melville K., Gadd G. M., McDermott F. P. (2005) Characterization of fungal community structure on a weathered pegmatitic granite. *Microbial Ecology*, Vol.0, pp.1–9. *ISSN* 0948-3055

Gollotte A., van Tuinen D., Atkinson D. (2004) Diversity of arbuscular mycorrhizal fungi colonising roots of the grass species *Agrostis capillaris* and *Lolium perenne* in a field experiment *Mycorrhiza*, Vol.14, pp.111–117. *ISSN*: 1432-1890. *ISSN*: 0940-6360

Gomez S.K. & Harrison M.J. (2009) Laser microdissection and its application to analyze gene expression in arbuscular mycorrhizal symbiosis. *Pest Manag. Sci.*, Vol. 65, pp.504–511. *ISSN*: 1526-498X

Guadarrama P., Álvarez-Sánchez F.J. (1999) Abundance of arbuscular mycorrhizal fungi spores in different environments in a tropical rain forest, Veracruz, Mexico. *Mycorrhiza*, Vol.8, pp.267–270. *ISSN*: 1432-1890. *ISSN*: 0940-6360

Harley J.L. & Harley E.L. (1987). A check-list of mycorrhiza in the British flora. *New Phytologist,* Vol.105, pp.1–102. *ISSN:* 1469-8137

Harner M.J., Opitz N., Geluso K., Tockner K., Rillig M.C. (2011) Arbuscular mycorrhizal fungi on developing islands within a dynamic river floodplain: an investigation across successional gradients and soil depth. *Aquat Sci,* Vol.73, pp.35–42. *ISSN:* 1015-1621

Harrison M.J. (1999). Molecular and cellular aspects of the arbuscular mycorrhizal symbiosis. *Annu. Rev. Plant. Physiol. Plant. Mol. Biol.,* Vol.50, pp.361-389. *ISSN:*1040-2519

Hausman N.T., Hawkes C.V. *(2009)* Plant neighborhood control of arbuscular mycorrhizal community composition. *New Phytologist,* Vol.183, No.4, pp.1188-1200. *ISSN:* 1469-8137

HawkswortH D.L. (2001) The magnitude of fungal diversity: the 1±5 million species estimate revisited *Mycol. Res.,* Vol.105, No.12, pp.1422-1432. *ISSN:* 0953-7562.

Helgason T., Daniell T.J., Husband R., Fitter A.H.,Young J.P.W. (1998). Ploughing up the wood-wide web? *Nature,* Vol. 384, pp.431. *ISSN* : 0028-0836

Helgason T., Fitter A.H. (2009) Natural selection and the evolutionary ecology of the arbuscular mycorrhizal fungi (Phylum Glomeromycota). *J. of Experimental Botany,* Vol.60, No.9, pp.2465-2480. *ISSN:* 0022-0957

Helgason T., Fitter A.H., Young J.P.W. (1999) Molecular diversity of arbuscular mycorrhizal fungi colonising *Hyacinthoides non-scripta* (bluebell) in a seminatural woodland. *Molecular Ecology,* Vol.8, pp.659-666. *ISSN:* 0962-1083

Helgason T., Merryweather J.W., Denison J., Wilson P., Young J.P.W., Fitter A.H. (2002). Selectivity and functional diversity in arbuscular mycorrhizas of co-occurring fungi and plants from a temperate deciduous woodland. *J. Ecol.* Vol.90, pp. 371-384. *ISSN:* 1365-2745.

Hibbett D.S.M., Binder J.F., Bischoff M., Blackwell P.F., Cannon O.E., Eriksson S., Huhndorf T., James P.M., Kirk R., Lücking T., Lumbsch F., Lutzoni P.B., Matheny D.J., Mclaughlin M.J., Powell S., Redhead C.L., Schoch J.W., Spatafora J.A., Stalpers R., Vilgalys M.C., Aime A., Aptroot R., Bauer D., Begerow G.L., Benny L.A., Castlebury P.W., Crous Y.-C., Dai W., Gams D.M., Geiser G.W., Griffith C., Gueıdan D.L., Hawksworth G., Hestmark K., Hosaka R.A., Humber K., Hyde J.E., Ironside U., Kõljalg C.P., Kurtzman K.-H., Larsson R., Lichtwardt J., Longcore J., Miądlikowska A., Miller J.-M., Moncalvo S., Mozley-Standridge F., Oberwinkler E., Parmasto V., Reeb J.D., Rogers C., Roux L., Ryvarden J.P., Sampaio A., Schüssler J., Sugiyama R.G., Thorn L., Tibell W.A., Untereiner C., Walker Z., Wang A., Weir M., Weiss M.M., White K., Winka Y., Yao J., Zhang N. (2007). A higher level of phylogenetic classification of the fungi. *Mycological Research,* Vol.111, pp.509–547. *ISSN:* 0953-7562

Hijri M. and Sanders I.R. (2004) The arbuscular mycorrhizal fungus *Glomus intraradices* is haploid and has a small genome size in the lower limit of eukaryotes. *Fungal Genetics and Biology,* Vol.41, pp.253–261. *ISSN:* 1087-1845

Hong J.W., Fomina M., Gadd G.M. (2009) F-RISA fungal clones as potential bioindicators of organic and metal contamination in soil. *Journal of Applied Microbiology,* Vol.109, pp.415-430. *ISSN:* 1365-2672

Horton T. R., Cazaras E., Bruns T. D. (1998) Ectomycorrhizal, vesicular-arbuscular and dark
    septate fungal colonization of bishop pine (*Pinus muricata*) seedlings in the first 5
    months of growth after wildfire. *Mycorrhiza,* Vol.7, pp.11-18. *ISSN*: 1432-1890

Husband R., Herre E.A., Turner S.L., Gallery R. and Young J.P.W. (2002a). Molecular
    diversity of arbuscular mycorrhizal fungi and patterns of host association over time
    and space in a tropical forest. *Mol. Ecol.* Vol.11, pp.2669-2678. *ISSN*: 0962-1083

Husband R., Herre E.A., Young J.P.W. (2002b). Temporal variation in the arbuscular
    mycorrhizal communities colonising seedlings in a tropical forest. *FEMS Microbiol.
    Ecol.* Vol.42, pp.131-136. *ISSN*: 0168-6496

Jansa J., Mozafar A., Banke S., McDonald B. A., Frossard E. (2002) Intra- and intersporal
    diversity of ITS rDNA sequences in *Glomus intraradices* assessed by cloning and
    sequencing, and by SSCP analysis *Mycol. Res.*, Vol.106, No.6, pp. 670-681.*ISSN*:
    0953-7562

Jefferies P., Gianinazzi S., Perotto S., Turnau K., Barea J.M. (2003). The contribution of
    arbuscular mycorrhizal fungi in sustainable maintenance of plant health and soil
    fertility. *Biol. Fert. Soils,* Vol.37, pp.1–16. *ISSN*: 0178-2762

Kapoor R., Sharma D., Bhatnagar A. K. (2008) Arbuscular mycorrhizae in micropropagation
    systems and their potential applications. *Scientia Horticulturae,* Vol.116, pp.227–239.
    ISSN: 0304-4238

Kjøller R., Rosendahl S. (2000) Detection of arbuscular mycorrhizal fungi (*Glomales*) in roots
    by nested PCR and SSCP (Single Stranded Conformation Polymorphism). *Plant and
    Soil,* Vol.226, pp.189–196. *ISSN*: 1573-5036

Klamer M., Roberts M. S., Levine L. H., Drake B. G., Garland J. L. (2002) Influence of
    elevated CO_2 on the fungal community in a coastal scrub Oak Forest soil
    investigated with terminal-restriction fragment length polymorphism analysis.
    Applied and Environmental Microbiology, Vol.68, No.9, pp.4370–4376. *ISSN*: 0099-
    2240.

Koch A.M., Croll D, Sanders I.R. (2006) Genetic variability in a population of arbuscular
    mycorrhizal fungi causes variation in plant growth. *Ecology Letters,* Vol.9, pp.103–
    110. *ISSN*: 1461-023X

Kowalchuk G. A., Gerards S., Woldendorp J. W. (1997) Detection and characterisation of
    fungal infections of *Ammophila arenaria* (marram grass) roots by denaturing
    gradient gel electrophoresis of specifically amplified 18S rDNA. *Applied and
    Environmental Microbiology,* Vol.63, pp.3858-3865. *ISSN*: 0099-2240

Kowalchuk G. A., Souza F. A., Van Veen J. A. (2002) Community analysis of arbuscular
    mycorrhizal fungi associated with *Ammophila arenaria* in Dutch coastal sand dunes.
    Molecular Ecology, Vol.11, pp.571-581. *ISSN*: 0962-1083

Krüger M., Stockinger H., Krüger C. and Schüßler A. (2009) DNA-based species level
    detection of *Glomeromycota*: one PCR primer set for all arbuscular mycorrhizal
    fungi. *New Phytologist,* Vol.183, pp.212–223. *ISSN*: 1469-8137

Lanfranco L., Delpero M., and Bonfante P. (1999) Intrasporal variability of ribosomal
    sequences in the endomycorrhizal fungus *Gigaspora margarita*. *Molecular Ecology,*
    Vol.8, pp.37–45. *ISSN*: 0962-1083

Lee J., Lee S., Young P. W. (2008) Improved PCR primers for the detectionand identication of
    arbuscular mycorrhizal fungi. *FEMS Microbiol Ecol,* Vol.65, pp.339-349. *ISSN*: 0168-
    6496.

Lekberg Y., Koide R.T., Rohr J.R., Aldrichwolfe L., Morton J.B. (2007) Role of niche restrictions and dispersal in the composition of arbuscular mycorrhizal fungal communities. *Journal of Ecology*, Vol.95, pp.95–105. *ISSN*: 1365-2656.

Lianga Z., Drijbera R. A., Leea D. J., Dwiekata I. M., Harrisb S. D., Wedin D. A. (2008) A DGGE-cloning method to characterize arbuscular mycorrhizal community structure in soil. *Soil Biology and Biochemistry*, Vol.40, pp.956–966. *ISSN*: 0038-0717

Lim S., Notley-McRobb L., Carter D.A. (2004) A comparison of the nature and abundance of microsatellites in 14 fungal genomes. *Fungal Genet. Biol.*, Vol.41, pp.1025-1036. *ISSN* 1087-1845

Liu W. T., Marsh T. L., Cheng H., Forney L. J. (1997) Characterization of microbial diversity by determining terminal restriction fragment length polymorphisms of genes encoding 16S rRNA. *Appl Environ Microbiol*, Vol.63, pp. 4516–4522. *ISSN*: 0099-2240

Long L.K., Yao Q., Guo J., Yang R.H., Huang Y.H., Zhu H.H. (2010) Molecular community analysis of arbuscular mycorrhizal fungi associated with five selected plant species from heavy metal polluted soils. *European Journal of Soil Biology*, Vol.46, pp.288-294.*ISSN*: 1164-5563

Longato S. & Bonfante P. (1997) Molecular identification of mycorrhizal fungi by direct amplification of microsatellite regions. *Mycol Res.*, Vol.101, No.4, pp.425-432. *ISSN*: 0953-7562

Luoma D. L., Eberhart J.L., Amaranthus M. P. (1997). Biodiversity of ectomycorrhizal types from Southwest Oregon. In: Kaye T. N., Liston A., Love R.M., Luoma D.L., Meinke R.J., Wilson M.V. (Eds.) Conservation and Management of Native Plants and Fungi. Native Plant Society of Oregon, Corvallis, OR, pp. 249–253.

Marjanovic´ Z., Uehlein N., Kaldenhoff R., Zwiazek J. J., Weiss M., Hampp R. D., Nehls U. (2005) Aquaporins in poplar: What a difference a symbiont makes! *Planta*, Vol.222, pp.258–268. *ISSN*: 0032-0935

Marsh T.L. (1999). Terminal restriction fragment length polymorphism (T-RFLP): an emerging method for characterizing diversity among homologous populations of amplification products. Curr Opin Microbiol., Vol.2, No.3, pp.323-327. *ISSN*: 1369-5274

Martin F., Gianinazzi-Pearson V., Hijri M., Lammers P., Requena N., Sanders I. R., Shachar-Hill Y., Shapiro H., Tuskan G. A., Young J. P. W. (2008) The long hard road to a completed *Glomus intraradices* genome. *New Phytologist*, Vol.180, pp.747–750. *ISSN*: 1469-8137

Martínez-García L. B., Armas C., Miranda J. D., Padilla F. M., Pugnaire F. I. (2011) Shrubs influence arbuscular mycorrhizal fungi communities in a semi-arid environment. *Soil Biology and Biochemistry*, Vol.43, pp.682-689. *ISSN*: 0038-0717.

Mathimaran N, Falquet L, Ineichen K, Picard C, Redecker D, Boller T, Wiemken A. (2008) Microsatellites for disentangling underground networks: strain-specific identification of *Glomus intraradices*, an arbuscular mycorrhizal fungus. *Fungal Genetics and Biology*, Vol.45, pp.812-817. *ISSN*: 1087-1845

Mergulhao A.C.D.S., da Silva M.V., Figueiredo M.D.B., Burity H.A., Maia L.C. (2008) Characterization and identification of arbuscular mycorrhizal fungi species by PCR/RFLP analysis of the rDNA internal transcribed spacer (ITS). *Annals of Microbiology*, Vol. 58, No.2, pp.341-344. *ISSN*: 1590-4261

Michelsen A., Rosendahl S. (1990) The effect of VA mycorrhizal fungi, phosphorus and drought stress on the growth of *Acacia nilotica* and *Leucaena leucocephala* seedlings. Plant and Soil, Vol.124, pp.7-13. *ISSN*: 1573-5036

Molina R., Massicotte H., Trappe J. M. (1992) *Specificity phenomena in mycorrhizal symbioses: community-ecological consequences and practical implications.* In: Mycorrhizal functioning: an integrative plant–fungal process (ed. M. F. Allen), pp. 357-423. New York: Chapman and Hall.

Moreira M., Baretta D., Tsai S.M., Cardoso E.J.B.N. (2009) Arbuscular mycorrhizal fungal communities in native and in replanted *Araucaria* forest. *Sci. Agric.*, Vol.66, No.5, pp.677-684. *ISSN* 0103-9016

Moreira M., Nogueira M.A., Tsai S.M., Gomes da Costa S., Cardoso E.J.B.N. (2007) Sporulation and diversity of arbuscular mycorrhizal fungiin Brazil Pine in the field and in the greenhouse. *Mycorrhiza*, Vol.17, pp.519-526. *ISSN*: 0940-6360

Mummey D. L., Antunes P. M., Rillig M. C. (2009) Arbuscular mycorrhizal fungi pre-inoculant identity determines community composition in roots. *Soil Biology and Biochemistry*, Vol.41, pp.1173-1179. *ISSN*: 0038-0717

Mummey D. L., Rillig M. C. (2008) Spatial characterization of arbuscular mycorrhizal fungal molecular diversity at the submetre scale in a temperate grassland. *FEMS Microbiology*, Vol.64, No.2, pp.260-270. ISSN:1574-6968

Mummey D. L., Rillig M. C., Holben W. E. (2005) Neighboring plant influences on arbuscular mycorrhizal fungal community composition as assessed by T-RFLP analysis. *Plant and Soil*, Vol.271, pp.83–90. . *ISSN*: 1573-5036

Muyzer G., De Waal I. E. C., Uitierlinden A. G. (1993) Profiling of complex microbial populations by denaturing gradient gel electrophoresis analysis of polymerase chain reaction-amplified genes coding for 16S rRNA. *Applied and Environmental Microbiology*, Vol. 59, No.3, pp.695-700. *ISSN*: 0099-2240

Newman E.I., Devoy C.L.N., Easen N.J., Fowles K.J. (1994). Plant species that can be linked by VA mycorrhizal fungi. *New Phytologist*, Vol.126, pp.691–693. *ISSN*: 1469-8137

Nilsson R. H., Ryberg M., Abarenkov K., Sjökvist E., Kristiansson E. (2009) The ITS regions a target for characterization of fungal communities using emerging sequencing technologies. *FEMS Microbiol Lett*, Vol.296, pp.97–101. ISSN:1574-6968

O'Dell T. E., Massicotte H. B., Trappe J. M. (1993) Root colonization of Lupinus latifolius Agardh. and *Pinus contorta* Dougl. by Phialocephala fortinii Wang and Wilcox. *New Phytologist*, Vol.124, pp.93-100. *ISSN*: 1469-8137

Oliveira C.A., Sá N.M.H., Gomes E.A., Marriel I.E., Scotti M.R., Guimarães C.T., Schaffert R.E., Alves V.M.C. (2009) Assessment of the mycorrhizal community in the rhizosphere of maize (*Zea mays* L.) genotypes contrasting for phosphorus efficiency in the acid savannas of Brazil using denaturing gradient gel electrophoresis (DGGE). *Applied Soil Ecology*, Vol.41, pp.249-258. *ISSN*: 0929-1393

Öpik M., Metsis M., Daniell T. J., Zobel M., Moora M. (2009) Large-scale parallel 454 sequencing reveals host ecological group specificity of arbuscular mycorrhizal fungi in a boreonemoral forest. *New Phytologist* Vol.184, pp.424–437. *ISSN*: 1469-8137

Öpik M., Moora M., Liira J., Kõljalg U., Zobel M., Sen R. (2003) Divergent arbuscular mycorrhizal fungal communities colonize roots of *Pulsatilla* spp. in boreal Scots pine forest and grassland soils *New Phytologist*, Vol.160, pp.581–593. *ISSN*: 1469-8137

Öpik M., Moora M., Liira J., Zobel M. (2006) Composition of root-colonizing arbuscular mycorrhizal fungal communities in different ecosystems around the globe. *Journal of Ecology*, Vol.94, pp.778–790. *ISSN*: 1365-2745

Orita M., Iwahana H., Kanazawa H., Hayashi K., Sekiya T. (1989) Detection of polymorphisms of human DNA by gel electrophoresis as single-stranded conformation polymorphisms. *Proc. Nat. Acad. Sci.*, Vol.86, pp. 2766–2770. *ISSN*: 0027-8424

Pagano M.C., Utidab M.K., Gomesb E.A., Marrielb I.E., Cabelloc M.N., Scotti M.R. (2011) Plant-type dependent changes in arbuscular mycorrhizal communities as soil quality indicator in semi-arid Brazil. *Ecological Indicators*, Vol.11, pp.643–650. *ISSN*: 1470-160X

Patreze C. M., Paulo E. N., Martinelli A. P., Cardoso E. J. B., Tsai S. M. (2009) Characterization of fungal soil communities by F-RISA and arbuscular mycorrhizal fungi from *Araucaria angustifolia* forest soils after replanting and wildfire disturbances. *Symbiosis*, Vol.48, pp.164–172. *ISSN*: 1878-7665

Pearson J. N., Jakobsen, I. (1993). Symbiotic exchange of carbon and phosphorus between cucumber and three arbuscular mycorrhizal fungi. *New Phytol.*, Vol.124, pp.481–488. *ISSN*: 1469-8137

Rabie G.H. (1998). Induction of fungal disease resistance in *Vicia faba* by dual inoculation with *Rhizobium leguminosarum* and vesicular–arbuscular mycorrhizal fungi. *Mycopathologia*, Vol.141, pp.159–166. *ISSN*:0027-5530

Ramos-Zapata J. A., Guadarrama P., Navarro-Alberto J. and Orellana R. (2011a) Arbuscular mycorrhizal propagules in soils from a tropical Forest and an abandoned cornfield in Quintana Roo, Mexico: visual comparison of most-probable-number estimates. *Mycorrhiza*, Vol. 21, pp.139–144. *ISSN*: 0940-6360

Ramos-Zapata J.A., Zapata-Trujilla R., Ortiz-Diaz J.J., Guadarrama P. (2011b) Arbuscular mycorrhizas in a tropical coastal dune system in Yucatan, Mexico. *Fungal Ecology*, Vol.4, No.4, pp.256–261. *ISSN*: 1754-5048

Ranjard L., Poly F., Lata J.-C, Mougel C., Thioulouse J., S. Nazaret (2001) Characterization of bacterial and fungal soil communities by automated ribosomal intergenic spacer analysis fingerprints: biological and methodological variability. *Appl. Environ. Microbiol.*, Vol.67, pp.4479–4487. *ISSN*: 0099-2240

Redecker D. (2000) Specific PCR primers to identify arbuscular mycorrhizal fungi within colonized roots. *Mycorrhiza*, Vol.10, pp.73–80. *ISSN*: 0940-6360

Renker C., Heinrichs J., Kaldorf M., Buscot F. (2003) Combining nested PCR and restriction digest of the internal transcribed spacer region to characterize arbuscular mycorrhizal fungi on roots from the field. *Mycorrhiza*, Vol.13, pp.191–198. *ISSN*: 0940-6360

Rocha F. S., Saggin Júnior O.J., da Silva E.M.R., de Lima W.L. (2006) Dependência e resposta de mudas de cedro a fungos micorrízicos arbusculares. *Pesq. agropec. Bras*, Vol.41, No.1, pp.77–84. *ISSN* 0100-204X

Rosendahl S. (2008) Communities, populations and individuals of arbuscular mycorrhizal fungi. *New Phytologist*, Vol.178, pp.253–266. *ISSN*: 1469-8137

Rosendahl S. & Stukenbrock E.H. (2004) Community structure of arbuscular mycorrhizal fungi in undisturbed vegetation revealed by analyses of LSU rDNA sequences. *Molecular Ecology*, Vol.13, pp.3179-3186. *ISSN*: 0962-1083

Sanders I. R., Alt M., Groppe K., Boller T., Wiemken A. (1995) Identification of ribosomal DNA polymorphisms among and within spores of the Glomales: application to studies on the genetic diversity of arbuscular mycorrhizal fungal communities. *New Phytol.*, Vol.130, pp.419-427. *ISSN*: 1469-8137

Schüssler A., Gehrig H., Schwarzott D. Walker C. (2001) Analysis of partial *Glomales* SSU rRNA gene sequences: implications for primer design and phylogeny. *Mycol. Res.*, Vol.105, No.1, pp.5-15. *ISSN*: 0953-7562

Schüssler A., Schwarzotta D., Walker C. (2001) A new fungal phlyum, the *Glomeromycota*: phylogeny and evolution. *Mycol. Res.*, Vol.105, No.12, pp.1413-1421. *ISSN*: 0953-7562

Schwarzott D. & Schüßler A. (2001) A simple and reliable method for SSU rRNA gene DNA extraction, amplification, and cloning from single AM fungal spores. *Mycorrhiza*, Vol.10, pp.203-207. *ISSN*: 0940-6360

Shepherd M., Nguyen L., Jones M.E., Nichols J.D., Carpenter F.L. (2007) A method for assessing arbuscular mycorrhizal fungi group distribution in tree roots by intergenic transcribed sequence variation. *Plant Soil*, Vol.290, pp.259-268.

Silva-Junior J. P. da; Cardoso E.J.B.N. (2006) Arbuscular mycorrhiza in cupuaçu and peach palm cultivated in agroforestry and monoculture systems in the Central Amazon region. *Pesquisa Agropecuaria Brasileira*, Vol.41, No.5, pp.819-825. *ISSN* 0100-204X

Simon L., Lalonde M., Bruns T.D. (1992) Specific amplification of 18S fungal ribosomal genes from VA endomycorrhizal fungi colonising roots. *Applied and Environmental Microbiology*, Vol.58, pp.291-295.

Siqueira J.O. (1994) Micorrizas. In: Araújo R.S., Hungaria M. (Eds.) Microrganismos de importância agrícola. Embrapa-CNPAF, Embrapa-CNPSo. Embrapa-SPI, Brasília, pp. 151-194.

Siqueira J.O., De Souza F.A., Cardoso E.J.B.N., Tsai S.M. (2010) Micorrizas: 30 anos de pesquisas no Brasil. Lavras: UFLA, 716pp.

Smith S.E. & Read D.J. (1997) *Mycorrhizal Symbiosis*, 2nd Ed. Academic Press, San Diego, U.S.A.

Smith S.E. & Read D.J. (2008) *Mycorrhizal Symbiosis*, 3rd ed. Academic Press, Amsterdam, Netherland.

Stockinger H., Krüger M., Schüssler A. (2010) DNA barcoding of arbuscular mycorrhizal fungi *New Phytologist*, Vol.187, pp.461-474. *ISSN*: 1469-8137

Stürmer S.L. & Siqueira J.O. (2011) Species richness and spore abundance of arbuscular mycorrhizal fungi across distinct land uses in Western Brazilian Amazon. *Mycorrhiza*, Vol.21, pp.255-267. *ISSN*: 0940-6360

Tautz D. (1989) Hypervariabflity of simple sequences as a general source for polymorphic DNA markers. *Nucl. Acids Res.* Vol.17, No.16, pp.6463-6471.*ISSN* 0305-1048

Tedersoo L., Nilsson R.H., Abarenkov K., Jairus T., Sadam A., Saar I., Bahram M., Bechem E., Chuyong G., Koljalg U. (2010) 454 Pyrosequencing and Sanger sequencing of tropical mycorrhizal fungi provide similar results but reveal substantial methodological biases. *New Phytologist*, Vol.188, pp.291-301. *ISSN*: 1469-8137

Tiedje J.M., Asuming-Brempong S., NuÈsslein K., Marsh T.L., Flynn S.J. (1999) Opening the black box of soil microbial diversity. *Applied Soil Ecology*, Vol.13, pp.109-122. *ISSN*: 0929-1393

Troeh Z.I., Loynachan T.E. (2003) Endomycorrhizal fungal survival in continuous corn, soybean, and fallow. *Agronomy Journal*, Vol.95, No.1, pp.224-230. *ISSN*: 0002-1962

Trouvelot S., van Tuinen D., Hijri M., Gianinazzi-Pearson V. (1999) Visualization of ribosomal DNA loci in spore interphasic nuclei of glomalean fungi by fluorescence *in situ* hybridization. *Mycorrhiza*, Vol.8, pp.203. *ISSN*: 0940-6360

Uibopuu A., Moora M., Saks U., Daniell T., Zobel M., Opik M. (2009) Differential effect of arbuscular mycorrhizal fungal communities from ecosystems along management gradient on the growth of forest understorey plant species. *Soil Biology and Biochemistry*, Vol.41, pp.2141–2146.

Unterseher M., Jumpponen A., ÖPik M., Tedersoo L., Moora M., Dormann C.F., Schnittler M. (2011) Species abundance distributions and richness estimations in fungal metagenomics – lessons learned from community ecology. *Molecular Ecology*, Vol.20, pp.275–285. *ISSN*: 0962-1083

van de Voorde T.F.J., van der Putten W.H., Gamper H.A., Gera Hol W. H., Bezemer T.M. (2010) Comparing arbuscular mycorrhizal communities of individual plants in a grassland biodiversity experiment. *New Phytologist*, Vol.186, pp. 746–754. *ISSN*: 1469-8137

van der Heijden M.G.A., Bardgett R.D., van Straalen N.M. (2008) The unseen majority: soil microbes as drivers of plant diversity and productivity in terrestrial ecosystems. *Ecol Lett*, Vol.11, pp.296–310. *ISSN*: 1461-0248

van der Heijden M.G.A., Klironomos J.N., Ursic M., Moutoglis P., Streitwolf-Engel R., Boller T., Wiemken A., Sanders I.R. (1998) Mycorrhizal fungal diversity determines plant biodiversity, ecosystem variability and productivity. *Nature*, Vol.396, pp.69-72. . *ISSN* : 0028-0836

van Diepen L.T.A., Lilleskov E.A., Pregitzer K.S. (2011) Simulated nitrogen deposition affects community structure of arbuscular mycorrhizal fungi in northern hardwood forests. *Molecular Ecology*, Vol.20, pp.799–811. *ISSN*: 0962-1083

van Tuinen D., Jacquot E., Zhao B., Gianinazzi-Pearson V. (1998b) Characterization of root colonization profiles by a microcosm community of arbuscular mycorrhizal fungi using 25S rDNA targeted nested PCR. *Mol. Ecol.* Vol.7, pp.879-887. *ISSN*: 0962-1083

van Tuinen D., Jacquot E., Zhao B., Gollotte A., Gianinazzi-Pearson (1998) Characterization of root colonization profiles by a microcosm community of arbuscular mycorrhizal fungi using 25S rDNA-targeted nested PCR. Molecular Ecology, Vol.7, pp.879-887. *ISSN*: 0962-1083

Vandenkoornhuyse P, Leyval C (1998) SSU rDNA sequencing and PCR-fingerprinting reveal genetic variation within *Glomus mosseae*. *Mycologia*, Vol.90, No.5, pp.791-797. *ISSN*: 0027-5514

Verbruggen E., Röling W.F.M., Gamper H.A., Kowalchuk G.A., Verhoef H.A., van der Heijden M.G.A. (2010) Positive effects of organic farming on below-ground mutualists: large-scale comparison of mycorrhizal fungal communities in agricultural soils. *New Phytologist*, Vol.186, pp.968–979. *ISSN*: 1469-8137

Wagg C., Pautler M., Massicotte H. B., Peterson R. L. (2008) The co-occurrence of ectomycorrhizal, arbuscular mycorrhizal, and dark septate fungi in seedlings of four members of the Pinaceae. *Mycorrhiza* , Vol.18, No.2, pp.103-110. *ISSN*: 0940-6360

Waldrop M.P., Zak D.R., Blackwood C.B., Curtis C.D., Tilman D. (2006) Resource availability controls fungal diversity across a plant diversity gradient. *Ecology Letters*, Vol.9, pp.1127-1135. . *ISSN*: 1461-0248

White T.J., Bruns T., Lee S., Taylor J.W. (1990) Amplification and direct sequencing of fungal ribosomal RNA genes for phylogenetics. In: Innis, M.A., Gelfand, D.H., Sninsky, J.J., White, T.J. (Eds.), PCR Protocols: A Guide to Methods and Applications. Academic Press, San Diego, California, pp.315–322.

Williams et al. (2011) Growth and competitiveness of the New Zealand tree species *Podocarpus cunninghamii* is reduced by ex-agricultural AMF but enhanced by forest AMF. *Soil Biology and Biochemistry*, Vol.43, pp.339-345. ISSN: 0038-0717

Wu B., Hogetsu T., Isobe K., Ishii R. (2007) Mycorrhiza Community structure of arbuscular mycorrhizal fungi in a primary successional volcanic desert on the southeast slope of Mount Fuji. *Mycorrhiza*, Vol.17, pp.495–506. *ISSN*: 0940-6360

Wu F., Dong M., Liu Y., Ma X., An L., Young J.P.W., Feng H. (2011) Effects of long-term fertilization on AM fungal community structure and Glomalin-related soil protein in the Loess Plateau of China. *Plant Soil*, Vol.342, pp.233–247. *ISSN*: 1573-5036

Zangaro W., Nishidate F.R., Vandresen J., Andrade G., Nogueira M.A. (2007) Root mycorrhizal colonization and plant responsiveness are related to root plasticity, soil fertility and successional status of native woody species in southern Brazil. *Journal of Tropical Ecology*, Vol.23, pp.53-62. *ISSN*: 0266-4674

Zangaro W., Nogueira M.A., Andrade G. (2009) *Arbuscular mycorrhizal fungi as biofertilizers for revegetation programs*. In RAI, M.K., (ed.) Current Advances in Fungal Biotechnology. The Haworth Press, Nova Deli. p. 1-28.

Zézé A., Hosny M, Gianinazzi-Pearson V., Dulieu H. (1996) Characterization of a highly repeated DNA sequence (SC1) from the arbuscular mycorrhizal fungus *Scutellospora castanea* and its use as a diagnostic probe *in planta*. *Appl. Environ. Microbiol.*, Vol.62, pp.2443–2448. *ISSN*: 0099-2240

Zézé et al. (1997) Intersporal genetic variation of *Gigaspora margarita*, a vesicular arbuscular mycorrhizal fungus, revealed by M13 minisatellite-primed PCR. *Applied and Environmental Microbiology*, Vol.63, No.2, pp.676-678. *ISSN*: 0099-2240

Zhang H., Tang M., Chen H., Tian Z., Xue Y., Feng Y. (2010) Communities of arbuscular mycorrhizal fungi and bacteria in the rhizosphere of *Caragana korshinkii* and *Hippophae rhamnoides* in Zhifanggou watershed. *Plant Soil*, Vol.326, pp.415–424. *ISSN*: 1573-5036

Zhou Z. and Hogetsu T. (2002) Subterranean community structure of ectomycorrhizal fungi under *Suillus grevillei* sporocarps in a *Larix kaempferi* forest. *New Phytologist*, Vol.154, pp.529–539. *ISSN*: 1469-8137

4

Arthropods and Nematodes: Functional Biodiversity in Forest Ecosystems

Pio Federico Roversi and Roberto Nannelli
CRA – Research Centre for Agrobiology and Pedology, Florence
Italy

1. Introduction

Despite the great diversity of habitats grouped under a single name, forests are ecosystems characterized by the dominance of trees, which condition not only the epigeal environment but also life in the soil. Unlike other ecosystems such as grasslands or annual agricultural crops, forests are well characterized by precise spatial structures. In fact, we can identify three main layers in all forests: a canopy layer of tree crowns, including not only green photosynthesizing organs but also branches of various sizes; a layer formed by the tree trunks; a layer including bushes and grasses, which can sometimes be missing when not enough light filters through the canopy. To these layers must be added the litter and soil, which houses the root systems. Other characteristic features of forests, in addition to their structural complexity, are the longevity of the plants, the peculiar microclimates and the presence of particular habitats not found outside of these biocoenoses, such as fallen trunks and tree hollows. Even in the case of woods managed with relatively rapid cycles to produce firewood, forests are particularly good examples of ecosystems organized into superimposed layers that allow the maximum use of the solar energy.

The biomass of forests is largely stored in the trees, with the following general distribution: ca. 2% in the leaves and almost 98% in trunks, branches and roots. In conditions of equilibrium between the arboreal vegetation and animal populations, saprophages, which use parts of plants and remains of organisms ending up in the litter, play a very important role. Indeed, the consumption of living tissues by other organisms included in the second trophic level is minimal, ranging between 0.1% and 2.5% of the net primary production. Any interruption of the mechanisms of demolition and degradation of organic substances would quickly lead to an accumulation of organic matter harmful to the operation of the forest system. For example, in Italian Apennine beech woods, the mean production of leaves reaching the litter is ca. 2.7 t/ha (Gregori & Miclaus, 1985), while for beech woods in southern Sweden it has been estimated at 5.7 t/ha. This mass of organic material reaching the ground would submerge the beech woods within several decades. In truth, however, the functioning of these woods would decline much earlier due to the lack of recycling of immobilized elements. If we turn our attention from temperate woods to tropical forests, we can see that the process would soon lead to the rapid collapse of the system. In fact, rapid decay of material reaching the ground (an estimated mean value of 20 t/ha is considered reliable) (Rodin & Bazilevich, 1967) is a *conditio sine qua non* for a system with great luxuriance of vegetation but rather poor and fragile soils.

The oldest inhabitants of this planet are for the most part trees. Yet, even without discussing the great patriarchs of the Earth, we must consider that growth cycles of over a century are the norm for many European forests. These rhythms were used for the oak forests that furnished the wood for the ships of the Most Serene Republic of Venice and even earlier for the Roman galleys. Vikings ships were constructed in Scandinavia with wood obtained from large old growth trees, especially oaks. The earliest forests can be identified on the basis of fossilized trees and forest soils from the mid-Devonian, between 400 and 350 million years ago. Starting from this period, we see an increasing number of remains of trees as well as a parallel exponential increase of the biodiversity of terrestrial zoocoenoses, with particular reference to the increased numbers of arthropod species (Retallack, 1997).

Forests are relatively stable ecosystems in which the tree component mediates the flows of organic matter by means of its long-term cycles. They support complex plant and microbial communities and represent, even in temperate climates, the largest storehouses of animal biodiversity, largely belonging to the Phyla Arthropoda and Nematoda. In table 1 are reported the values estimated in abundance of arthropod fauna living in a tropical forest (Seram rain forest).

Biotopes	N (millions/ha)
soil	23.7
leaf litter	6.0
ground vegetation	0.1
tree trunks	0.5
canopy	12.0
Total	42.3

Table 1. Abundance of arthropod fauna estimated in a tropical forest (Stork,1988).

Present-day forests host a diversified and largely specific fauna able to use not only the trees but also bushes and grasses. Trees are also more or less covered by microalgae, epiphytic lichens and mosses which can serve as food for phytophages, estimated to make up 20% of the total number of species in Danish beech woods (Nielsen, 1975). The primary production of epiphytes in many temperate forests is comparable to the production of the herbaceous layer, and some groups of arthropods specifically associated with these food substrates, such as Psocidae, can reach a density of 4,000 individuals per square metre of bark (Turner, 1975).

Moreover, the levels of biodiversity can vary enormously among different types of woods and within the same forest according to the age of the tree populations or the spatial heterogeneity characterizing woods of different ages. There are also trees with an extremely reduced arthropod fauna due to the production of phytoecdysones similar to moulting hormones, which act as repellents. In contrast, other trees host a much higher number of primary consumers due to the presence in their tissues of salicylic acid derivates, which act as attractants.

It is important to underline that when we speak of biodiversity we are often referring to three principal components:

-   genetic diversity;
-   diversity of species;
-   diversity of habitats or ecosystems.

However, a fourth component was recently proposed, namely "FUNCTIONAL DIVERSITY", based on the trophic role of the species and the interactions among organisms and between them and their environment.

In forest ecosystems, the animal component is distributed differently in the various layers, playing a fundamental role in terms of the following functional groups:

-   primary consumers or herbivores;
-   demolishers;
-   degraders (often associated with symbiotic micro-organisms).

Arthropods in the different functional groups can be further divided into ecological-nutritional subgroups or categories based on their feeding behaviour:

-   leaf suckers - species that puncture plant organs or tissues and suck the cell contents or substances circulating in the trees;
-   defoliators - species that feed on green parts of the trees (leaves, buds and unlignified bark) throughout the life cycle or only in some stages;
-   xylophagous - species able to feed throughout their life cycle or for part of it on lignified tree structures Xylophagous are further subdivided into:
    -   corticolous,
    -   corticolous-lignicolous,
    -   lignicolous.

2. Tree crown

Arthropods that live in tree crowns present a great variety of biological aspects. They are represented mainly by insects, spiders and mites, and to a lesser extent by representatives of other groups such as Apterygota, Chilopoda and Diplopoda. Not all of these animals live stably in the canopy, with various species using the crown only for part of their biological cycle (various lepidopterans feed on leaves but when their larval development is completed they move into the soil to undergo metamorphosis, e.g. *Lycia hirtaria* (Clerck) in European temperate forest and *Erannis tiliaria* (Harris) in Canadian mixedwood boreal forest. In contrast, other arthropods, such as coleopterans of the genus *Polydrusus*, only feed on leaves and needles as adults after living in the soil as larvae and feeding on roots.

Crown arthropods can be grouped as follows:

-   phyllophages, which feed on leaves or needles, eroding them from the outside (mainly lepidopterans plus some hymenopterans and coleopterans);
-   leaf and bud miners (above all lepidopterans plus some coleopterans);
-   floricolous arthropods:
    -   anthophages,
    -   pollenophages;

- arthropods that develop by eating fruits (carpophages) and seeds (spermophages);
- arthropods that induce the formation of galls, with shapes that vary according to the species involved (cinipid hymenopterans, dipterans and mites);
- suckers of cell contents and sap (mainly but not exclusively elaborated sap – above all Rhyncota insects of the suborder Homoptera);
- herbivores that feed on algae (algophages), fungi (mycetophages) and lichens (lichenophages) (mainly psocopterans);
- detritivores (including dermapterans, psocopterans,...);
- predators (many coleopterans but also some arboricolous orthopterans, Heteroptera Miridae, etc.);
- parasitoids (primarily hymenopterans but also many dipterans).

Primary consumers prevail in the canopy layer, especially phytophages in the ecological-nutritional categories of defoliators (with a masticatory buccal apparatus for feeding on leaves and green buds) and leaf suckers (provided with a puncturing-sucking buccal apparatus). A study conducted on oaks, black locusts and birches in England and South Africa showed substantial similarity of the crown arthropod communities between the two areas, with phytophages representing ca. one quarter of all species, half of the biomass and two thirds of all individuals (Moran & Southwood, 1982).

Defoliator insects are common in all coniferous and broadleaf forest ecosystems. They are an integral part of the forests and they can be useful to the ecosystem or harmful to the conservation and productivity of these habitats according to the quantity of photosynthesizing material removed and the possible repetition of severe attacks. In many forest formations, a primary role is played by various families of defoliator lepidopterans. Formations purely or prevalently of oaks provide a particularly important example. In Italy, there is a wide variety of biogeographical contexts within a relatively small surface area, and this is reflected in the number of *Quercus* species. The middle European oaks, often distributed in a fragmentary manner on plain and hilly terrains now largely dominated by agriculture, are joined by the larger contingent of Mediterranean species, the most important being the cork oak and holm oak. These oaks are very important in helping to form some of the most complex forest ecosystems in Italy, with an extremely various phytophagous arthropod fauna able to utilize the different ecological niches available in relation to seasonal conditions, composition, age and silviculture treatments. This heterogeneous complex of wooded formations, where human activities have profoundly modified the phytocoenoses, supports a rich and diversified lepidopteran fauna. Over 200 lepidopteran species belonging to 32 families feed on the green parts of the crowns of these Fagaceae, eroding buds, leaves and shoots from the outside or mining into them (Luciano and Roversi, 2001; Cao and Luciano, 2007). They constitute the most numerous group of phytophagous insects. The pre-eminence of this group is related to bio-ethological characteristics common to most of the species:

- the more or less strong polyphagy, which allows them to best utilize the occasionally available food;
- the possibility of laying eggs on different substrates without being bound to the trees that will serve as food for the larvae;
- the great variability of biological cycles, with species that pass the winter in different developmental stages;
- the absence of diapause (with rare exceptions), allowing the species to express the maximum potential for numerical increase in each generation;

- amphigonic reproduction, which provides the species with greater adaptability;
- the ability to carry out migrations or at least medium- and long-term movements, favouring rapid diffusion into suitable territories;
- possibilities for passive diffusion by some species, particularly in the first larval stages.

Lepidoptera are also cited as major defoliators of tropical forest trees and there are actually an increasing number of publications dedicated to the attack in neotropical forest environments (Nair, 2007; Haugaasen, 2009).

Thanks to these characteristics, a small group of oak-defoliating lepidopterans can give rise to spectacular numerical explosions or "outbreaks" with the sudden defoliation of whole stands, as in the case of mass *Lymantria dispar* (L.) infestations (Fig. 1). From an ecological point of view, outbreaks present characteristic phases of population abundance.

Fig. 1. *Lymantria dispar* larvae on *Quercus pubescens*.

After remaining at low numerical levels for years, becoming almost impossible to observe (latency phase), the phytophage populations show rapid density increases (progradation phase) which in several years lead to true numerical explosions or mass infestations (peak phase), followed by a more or less rapid decline (retrogradation phase) toward a new latency phase. In some defoliator lepidopterans, these mass changes are fairly common and are regulated by various biotic and abiotic factors that strongly affect the vitality and survival of the individuals. The main factors include climate (able to influence populations directly, e.g. allowing a higher or lower number of individuals to survive winter, and indirectly, by modifying the phenology of the host trees and thus the quantity and quality of food available for species that feed on the trees at the beginning of spring) and the set of natural antagonists (predators, parasitoids, pathogens) (Fig.2).

Fig. 2. *Calosoma sycophanta* a voracious Carabid consumer of caterpillars.

The different combinations of these factors produce more or less intense demographic variations, which allow species with a high biotic potential to reach high enough densities to cause the complete removal of the leaf mass of entire forests, even extending over hundreds of thousands of hectares. These mass appearances can occur at regular intervals or give rise to attacks irregularly distributed in space and time.

Very dramatic outbreaks often occur in degraded forest ecosystems or in marginal zones of the distribution area of the tree species where the appearances of defoliators are more frequent. This aggravates existing ecological imbalances, leading to very severe situations of widespread tree weakening, predisposing them to subsequent attacks by aggressors able to cause an irreversible decline. During such outbreaks, a large part of the organic matter produced by the trees can be diverted from the normal circuit. A good example is the activity of *Operophthera brumata* (L.), a geometrid able to defoliate entire stands in Swedish forests, whereas in the latency period the larvae remove no more than 0.004÷0.006 of the net primary production (Axelsson et al., 1975).

Leaf sucking insects have a puncturing-sucking buccal apparatus in which the mandibles and maxillae are profoundly modified to form an efficient tool similar to a hypodermic needle. They use it to suck sap and cell contents from the trees, becoming the first factor of deviation of part of the organic matter otherwise destined for the production of photosynthesizing and reproductive organs, the constitution of reserves and the accumulation of wood mass.

Three large groups of leaf sucking insects can be identified: species that feed by directly puncturing cells or groups of cells in superficial tissues (as in the case of thysanopterans) or deep tissues (as for some Homoptera Rhyncota); species that feed on elaborated sap, which stick their buccal stylets directly into the phloem conducting system (including Homoptera Rhyncota, mainly Aphidoidea, Coccoidea, Psylloidea and Aleyrodoidea); species that reach the xylem transport pathways, feeding on crude sap (including a small group of homopterans such as some Cercopidae). In these last species, also known as "spit bugs", the larval stages are found inside large foamy masses formed by the insects to eliminate the excess of water they are forced to ingest: indeed, the low concentration of nitrogen compounds in crude sap compels the Cercopidae to ingest large quantities of food in order to extract the necessary amounts of nutrients (Thompson, 1994). Adults of Cercopidae, including species that complete their larval development in the soil, also withdraw food and liquids by tapping into the xylem circuit of the leaves of host plants by means of refined behavioural adaptations: for instance, adults of *Haematoloma dorsatum* (Ahrens) puncture the needles of pines and other conifers by introducing their buccal stylets through the stomata to reach the feeding sites without an excessive energy expenditure (Roversi et al., 1990).

Leaf suckers that feed on elaborated sap consume large quantities of sugars which are not completely used and thus are expelled as drops of honeydew, highly attractive to ants, bees, wasps and adults of predators and parasitoids of aphids. Hence, these animals consume a part of the products of photosynthesis by trees that are otherwise unavailable to them. Therefore, in conditions of general equilibrium of forests, leaf suckers (particularly those that produce honeydew) are integral parts of the system, allowing an intensification of the network of interrelationships within the ecosystem and between it and other biocoenoses.

With the changing of the seasons, trees are subjected to strong variability in the transfer of nutrients and mineral salts. This is extremely important for leaf suckers, which gain their nutriment from the host tree's vascular system. In Italian environments, there can be strong differences between deciduous and evergreen trees, since the former are subjected not only to a marked translocation of crude and elaborated materials during growth resumption in spring but also to an autumn re-allocation of nitrogen substances and others when the leaves fall.

Tree crowns also host arthropods with such close associations with trees that they modify the host's local physiological processes and morphology, forcing them to create particular structures, called galls, in which the arthropods develop and shelter. There is a great morphological variety of galls, with forms and colours characteristic of each insect species. This is particularly true for structures formed by Hymenoptera Cynipidae and Diptera Cecidomyiidae, which have long attracted the attention of students of nature, starting with Francesco Redi who in the 17th century dedicated three manuscripts to them, including tempera drawings of the insects and their larvae (Bernardi et al., 1997). In some habitats such as beech woods, the unmistakable reddish leaf galls of the cecidomyiid *Mikiola fagi*, with their larval contents, are a food supplement for micromammals in an ecosystem in which food reserves for these vertebrates are particularly scarce.

On the crowns of both coniferous (various *Pinus* spp.) and broadleaf trees (oaks), we find insects that carry out most of their life cycle elsewhere, usually in subcortical and woody tissues, but which require a relatively brief period of feeding on vigorous shoots for maturation of their reproductive apparatus. They include Coleoptera Scolytidae of the genera *Tomicus* and *Scolytus*, which burrow galleries in the shoots, and Cerambycidae of the genus *Monochamus*, which erode young bark, all widely distributed in temperate forests. The alternation of feeding sites between woody tissues of trees often weakened by pathogenic infections and peripheral parts of the crown introduces another peculiar aspect into the relationships among very different groups of animals. Scolytidae and Cerambycidae with this behaviour can act as vectors, not only carrying fungal spores directly into contact with living tissues of new host trees but also becoming efficient vehicles of inoculation of healthy plants with phytopathogenic nematodes, which benefit from the damage caused by the xylophages to reach the conductor system of the host trees. Among the various relationships between nematodes and arthropods, especially insects, the associations established by the phoretic entomophilous nematodes of the genus *Bursaphelenchus* can have very severe consequences. In favourable conditions, they can allow the nematodes, transported by beetles under their elytra or in the trachea, to multiply with extreme rapidity in the conducting tissues of the trees (Covassi & Palmisano, 1997).

Acari living in the forest canopy (on leaves and small and main branches) are represented by various groups with different densities. Phytophagous mites belonging to the Tetranychidae, Tenuipalpidae and Eriophyidae, feed on the tree sap or cell contents. Their behaviour and densities are influenced by the environmental conditions and by the cohort of their predators, mainly Acari Mesostigmata and Prostigmata (Figs. 3 and 4) of the families Phytoseiidae, Stigmaeidae, Cheyletidae, Anystidae, Cunaxidae, Bdellidae and Trombidiidae. Their behaviour is also influenced by the type of plant, i.e. broadleaf or evergreen. Phytophagous species living on deciduous trees in unfavourable periods tend to move onto the branches and trunk to find micro-environments suitable for overwintering. On coniferous and evergreen trees, which keep their leaves even in winter, phytophagous mites tend to shelter at the base of the leaf nervation or in cracks in small branches.

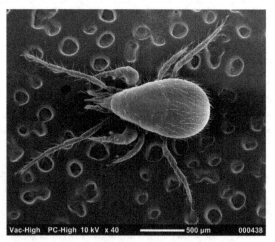

Fig. 3. A predatory mesostigmatid mite living in the forest canopy (SEM x 40).

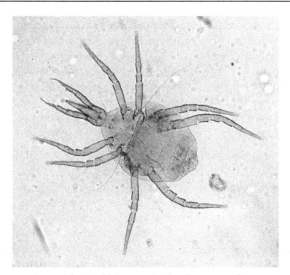

Fig. 4. A predatory prostigmatid mite living on bark of tree trunks.

Tree crowns host a wide variety of Acari: fungivores, algophages or those associated with mosses and lichens. These Acari belong to different groups, including Oribatida or Cryptostigmata, Astigmata and Prostigmata Tydeidae and Tarsonemidae, in fine equilibrium with their predators, generally Acari Mesostigmata and other Prostigmata, which habitually frequent leaves and small branches.

3. Tree trunks

The bark of tree trunks is an important element of forest ecosystems. It creates particular microclimates and provides micro-habitats that differ greatly according to the rugosity and irregularity of the surface, particularly in terms of the presence of deep fissures. In this spatial sector, which hosts phytophages as well as various predators and detritivores, the dominant groups are spiders, mites and insects, mainly springtails, psocids and some coleopterans and dipterans (Funke & Summer, 1980).

The trunk can contain mites associated with mosses and lichens but also strictly corticolous species, which find food and protection there. Galleries burrowed under the bark by scolytid, cerambycid and buprestid beetles host Acari belonging to Astigmata, Mesostigmata and particularly Prostigmata Tarsonemidae, which feed on the fungi typical of this environment. Other mites live in rifts in the bark caused by cortical cancers, feeding on fungal mycelia and fruiting bodies or on the associated microflora.

Also active on trunks are leaf suckers able to reach the tree's lymphatic system with long buccal stylets, e.g. the highly mimetic Heteroptera Rhyncota of the genus *Aradus* and cochineals of the genus *Matsucoccus*. Trunk-living insects with a puncturing-sucking buccal apparatus also include predators with a high degree of adaptation; this is observed especially in species of the genus *Elatophilus*, whose body is very flattened and admirably adapted to working its way between bark scales (Covassi & Poggesi, 1986).

Bark and especially wood constitute the main reserves of organic matter and mineral elements, with values that can vary enormously according to the vegetational composition, age of the woods and their conservation state. A beech wood of the Fontainbleau Forest in France had an estimated total tree biomass of over 280 t/ha, of which 195 represented by trunks, 48 by branches, 41 by roots, 3.2 by leaves and sprigs, and 0.9 by reproductive organs (Dajoz, 2000). In the so-called "Old Forests" of the western United States, consisting of coniferous associations with a mean age of 450 years, the epigeal biomass varied from 492 to 976 t/ha, with woody residues varying between 143 and 215 t/ha (Spies et al., 1988).

However, bark and wood tissues are degraded much more slowly than leaves. Arthropods, and xylophages in particular, play a primary role in accelerating the processes of decomposition and mineralization, often acting simultaneously with various fungi. In a transverse section of a trunk with, from external to internal, bark or rhytidome, phloem, cambium and xylem or real wood (separated into sapwood and heartwood), we can see a progressive change in the composition in terms of cellulose, hemicellulose, lignin, pectin, suberin, starch, nitrogen compounds, lipids and minerals, with a clear prevalence of the first three compounds. Cellulose, a polymer formed by the association of a high number of carbohydrate molecules in both the amorphous form and crystalline form, constitutes from 40 to over 60% of the dry weight of wood and can be hydrolyzed only by fungi, bacteria and some isopods, diplopods and insects provided with cellulase.

Only a few insect species can completely degrade cellulose since many xylophages that feed on bark, subcortical tissues or wood have only one of the three enzymes necessary for the final conversion into glucose. Among the few arthropods able to totally degrade cellulose are Thysanura (Lasker & Giese, 1956), Dermaptera (Cleveland, 1934), Isoptera, Hymenoptera Siricidae, Coleoptera Anobiidae, Buprestidae, Cerambycidae and Scarabeidae. However, occasionally in the wild, some xylophagous species can ingest bacteria, fungi and protozoa, using them as a source of enzymes to increase their ability to digest cellulose, hemicellulose and lignin. In the case of Hymenoptera Siricidae that develop in the wood of coniferous trees, their ability to utilize this food depends on the fact that the larvae ingest wood invaded by fungi of the genus *Amylostereum* and establish a mutualistic symbiosis with them, which provides the digestive enzymes necessary to use the pabulum on which the larvae develop (Kukor & Martin, 1983).

Lignin forms 18-38% of the durable tissues of trees. However, this term groups very different complex substances into a single category and these substances make the degradation of wood even more difficult since no animals, including insects, produce enzymes able to degrade them. Micro-organisms that can decompose lignin include fungi, bacteria and protozoa, many of which live in the digestive apparatus of arthropods, establishing mutualistic symbioses that allow the hosts to degrade lignin.

The bark of tree trunks also constitute "highways" for the daily (ants of the genus *Rufa*) or seasonal movements of species that migrate from the soil or litter to the tree crown during their cycles. Many species use the bark of the trunk as mating, egg-laying or wintering sites. In some cases, the bark becomes a summer refuge for species that live on the crown but abandon it when the temperatures reach their summer peak, as in the case of the woolly oak aphid (Binazzi & Roversi, 1990).

The subcortical layers are the site of development of various species, mainly insects and especially scolytid and cerambycid beetles. The Scolytidae are one of the most important groups of phytophagous insects, linked mainly to forest habitats and particularly coniferous woods (Figs. 5 and 6). There they express their maximal diversity of cycles and modes of development, behaviours and abilities to settle on trees in different vegetative conditions, often becoming dominant elements in the mechanism regulating ecological successions (Chararas, 1962; Pennacchio et al., 2006). In Europe, 39 species have been recorded on *Pinus* alone. The family Scolytidae numbers over 6,000 species of small insects, usually of more or less cylindrical shape and less than 2 mm long, with mandibles that move horizontally and the initial part of the intestine provided with sclerified denticles that act as a filter preventing the ingestion of excessively large wood fragments. The family includes polygamous species, such as those in the genera *Ips*, *Orthotomicus*, *Pityophthorus*, *Pityogenes* and *Pityokteines*, or monogamous ones such as those in the genera *Tomicus* and *Cryphalus*, which burrow characteristic systems of breeding galleries where they lay their eggs. The galleries of adults and those dug by the larvae to complete their development are well determined for each species; indeed, it is possible to make an *a posteriori* identification of scolytid species that have developed on a given tree merely by examination of the damage.

Very many other insects, mites and non-hexapod invertebrates with diversified alimentary regimes are found in the subcortical galleries burrowed by Scolytidae. They contribute substantially to the maintenance of high levels of biodiversity of the animal communities living under tree bark.

Fig. 5. Galleries on Pine produced by *Tomicus minor*.

Fig. 6. Galleries in Spruce produced by *Ips typographus*.

Most scolytids attack trees with integral phloem and cambium tissues. However, these trees are often weakened for various reasons, including drought and attacks by other phytophages, primarily defoliators but also a small number of species that colonize trees in more or less advanced stages of decay with active fermentation processes in the subcortical tissues and with low levels of starch and protein contents, e.g. species in the genera *Dryocoetes* and *Hylurgops*.

It is important to underline that, even in scolytid species that attack only a single host, not all the trees are equally attractive. These small hexapods are refined aggressors able to perceive with their sensory organs changes in the spectrum of substances emitted by each individual tree as a result of biotic and abiotic stresses. The manner in which thousands of individuals of a given scolytid species swarm onto single trees among thousands of other unaffected trees illustrates the precision of these beetles' stimulus perception mechanism.

The relationships among trees, semiochemicals and xylophagous insects constitute one of the most interesting research fields in forest entomology, not only for an understanding of the subtle mechanisms that allow their correct functioning but also for their practical implications. The use of semiochemicals for the monitoring and mass capture of these phytophages is a current strategy for the management of biotic adversities in many forest areas because of the possibility of limiting beetle population increases via agents with low environmental impact and without the use of synthetic biocides.

Examples of species well known for their ability to modify the structure and species composition of wooded formations are various *Dendroctonus* in North America and *Ips typographus* in Europe. The latter species caused the loss of over 5 million cubic metres of timber in Norway in the period 1970-1982 alone. In the Palaearctic Region, *Scolytus* species on elms have been responsible, in association with the fungal agent of Dutch elm disease *Ophiostoma novo-ulmi* Brasier, for the devastation of elm trees, while *Tomicus* species have been a constant threat to pine woods along coastlines and on hills and mountains.

The Cerambycidae are medium to large-sized beetles with a lengthened body and generally long antennae, very common in forests since the larvae of most species develop under bark or in wood. They are important both for conservation and for maintenance of the functionality of the woods; although some species cause damage, others actively contribute to the degradation of dead wood, intervening effectively in the first phase of demolition of durable tree structures. Species of the genus *Monochamus*, particularly *Monochamus galloprovincialis galloprovincialis* (Olivier) in Mediterranean pine woods, are able to rapidly kill trees that are only momentarily weakened (Francardi & Pennacchio, 1996). The harmfulness of these and other beetles of the family is also due to their role as vectors of phytopathogenic nematodes of the genus *Bursaphelenchus* (Fig.7).

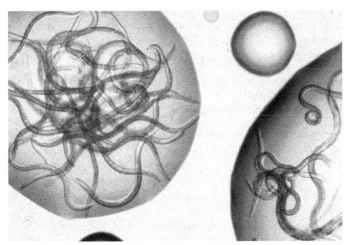

Fig. 7. *Bursaphelenchus eremus*: a phytoparasitic nematode living on *Quercus* spp.

Some lepidopterans develop by feeding on durable tree tissues, such as species of the genus *Dioryctria*. They are known as "pine resin moths", even though they can also live on cedars and red firs. Their larvae dig subcortical galleries exclusively on conifers thanks to their ability to isolate themselves from the abundant resin emitted by the trees in an attempt to defend themselves. To do so, they create tunnels lined with sericeous tissue.

Isopterans, various coleopteran families (including Scolytidae, Cerambycidae and Buprestidae), Lepidoptera Cossidae and Sesiidae, Hymenoptera Siricidae and some dipterans contain xylophagous species able to develop for part of their life or for the entire larval period by feeding on living wood or wood at various stages of decay. In the first case, we are referring to species included in the corticolous-lignicolous group whose larvae feed first on the phloem and later on the xylem, moving deep inside trunks and branches to complete their development and undergo metamorphosis. The second case involves xylophagous species *sensu stretto* or lignicolous ones that feed exclusively on woody tissues. They are fundamental in accelerating the decay of tree branches and trunks, and the group includes some of the largest of all arthropods. For instance, among Cerambycidae, the larvae of *Ergates faber* found on pines, oaks, chestnuts, elms, cherries and other hosts in Italy reach 8 cm in length, while the larvae of *Acrocinus longimanus*, distributed from Mexico to South America, are up to 13 cm long. The larval activity of these true burrowers can demolish age-old trees, even reaching the most resistant portions of the heartwood. Some xylophages *s.s.* are found on only one host species, although many develop by feeding on different trees; some species, mainly linked to decayed wood already invaded by fungi, are able to accept hundreds of coniferous and broadleaf species. On account of the poor pabulum, many species in this group, particularly those in relatively dry habitats, require several years to complete their larval development, even up to six years. Examples are the cerambycid *Hylotrupes bajulus* L. and a scarabaeid that lives in dead wood and is one of the best known and most characteristic of insects, the stag beetle (*Lucanus cervus* L.), whose adult males are provided with unmistakable long but absolutely harmless mandibles. Some xylophages found in dead wood in humid tree hollows are among the rarest of protected insects worldwide, e.g. the Scarabaeidae Dynastinae *Osmoderma eremita*. Various other coleopteran families include important xylophagous species, e.g. Buprestidae, Melandryidae (*Serropalpus barbatus*), Anthribidae and Tenebrionidae. The Diptera families Tipulidae, Cecidomyiidae and Asilidae contain species often found in dead wood.

Therefore, many xylophagous insects are important components of forest ecosystems, and, in conditions of equilibrium with the host trees, help maintain the vigour of the system, assuring the dynamism of the biocoenosis. In fact, forest insects and pathogens:

1.   colonize trees suitable for their development, modifying the structure and composition of the woods and modelling the tree population to the conditions offered by the site.
2.   participate in tree turnover via elimination of the less fit trees, accelerating the availability of elements and other substances to the healthier trees. Indeed, the action of corticolous, corticolous-lignicolous and lignicolous insects contributes to the decay of the durable tree parts, thus favouring the action of nitrogen-fixing bacteria and saprophytic micro-organisms which mineralize and liberate elements for roots and mycorrhizae.
3.   create peculiar habitats, via deformations of the crown, the death of some trees and the opening of clearings, and make available nutrients for other species that play a critical role in the maintenance of forest stability and productivity. A good example is the formation of suitable habitats and refuges for insectivorous birds.

Hence, the action of xylophagous insects and associated pathogenic agents plays a primary role in the maintenance of productivity and biodiversity levels in forest environments.

Fig. 8. Adult of *Ergates faber*.

4. Litter and soil

Although leaf-sucking and defoliator arthropods prevail in the crown layer, demolishers are of great importance in the lower layers and especially the litter, which receives not only leaves but also large amounts of organic matter in the form of fallen trunks and branches. Indeed, their role becomes essential in the phase of breakdown and decay of organic substances (Fig.9). To underline the complexity of the trophic networks, we should also mention that nematodes form an important part of all the functional groups in the soil, from phytoparasites active on root organs to bacteriophages with densities that can reach 30 million individuals/m² in forests.

The forest litter and soil are the environment with the greatest animal biodiversity and the site of recycling of materials and nutrients. The dominant groups of soil organisms, in terms of number and biomass, are micro-organisms such as bacteria, fungi and yeasts, representing ca. 85% of the biomass. However, other extremely important and complex components are represented by animals with sometimes very diverse dietary regimes, such as protozoa (amoebas, flagellates, ciliates), nematodes (bacteriophages, fungivores, omnivores, predators), Enchytraeidae, Lumbricidae and micro-arthropods, mainly mites (bacteriophages, fungivores, predators), springtails (fungivores and predators), and dipteran and coleopteran larvae, which make up the remaining 15% of the biomass. Nevertheless, the systematic spectrum of large taxonomic groups is limited with respect to the specific variety found within the same groups (Fig.10).

Table 2 summarize the soil fauna living in a temperate forest.

Fig. 9. *Abies alba* windfall in the Vallombrosa Forest (Central Italy).

Zoological groups	Number of individuals/m²	Weight in g
Protozoa	100 to 1000 x 10^6	2 to 20
Nematoda	1 to 30 x 10^6	1 to 30
Lumbricidae	50 to 400	20 to 250
Enchytraeidae	10 to 50 x 10^3	1 to 6
Acari	20 to 500 x 10^3	0.2 to 5
Pseudoscorpionida+ Araneidae+Opilionidae	60	0.06
Collembola	20 to 500 x 10^3	0.5 to 5
Protura	200	insignificant
Diplura	150	insignificant
Thysanura	few individuals	insignificant
Formicidae	variable according to site	-
Coleoptera larvae	100	1
Diptera larvae	400	3.5
Symphyla	1000	0.1
Chilopoda	50	1
Diplopoda	1 - 200	8
Isopoda	100	4

Table 2. Division of the forest soil fauna in temperate regions (values estimated from data of different authors). (From Bachelier, 1971).

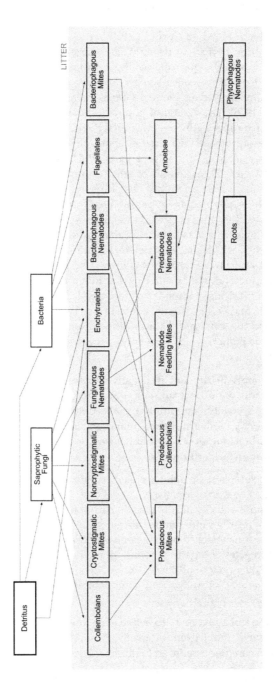

Fig. 10. Diagram of the food web in the forest floor and soil.

An important subdivision of soil-living animals is based on their size. This parameter is of primary importance in the edaphic environment since the organisms must move through the soil to search for food, breed and avoid natural antagonists and momentarily adverse conditions, as occur during periods of drought. The classification of soil animals based on size allows us to distinguish the macrofauna and microfauna from the mesofauna, which includes the major part of soil organisms and whole groups such as Diplura, Collembola and Protura as well as the majority of species of Tardigrada, Rotifera, Isopoda and especially Nematoda (Wallwork, 1976). The biological population of the soil has a dual origin: on the one hand a contingent of species living in capillary and gravitational water that fills or covers the soil cavities (protozoa, nematodes, most oligochaetes); on the other hand a different group originating from the epigeal part, which has adapted to life in the soil (almost all arthropods with the exception of small groups such as copepods). However, there is a third contingent, whose origin has been directly identified in Palaeozoic deposits, containing forms strongly adapted to hypogeal life (Insecta Symphyla, Collembola and Pauropoda, Protura and Diplura). The abundance and composition of these groups depend on the type of soil and overlying vegetation. Coniferous needles have a high C/N ratio and high polyphenol contents, which make this material more resistant to decomposition than the leaves in broadleaf litter, where the C/N is lower and decomposition is more rapid. In this type of litter, the mineral soil is much less acidic and the more active decomposer groups are favoured (lumbricids, millipedes and isopods). Soil arthropods are generally not uniformly distributed and often show aggregative behaviour, being distributed in horizontal aggregates in the same profile or in different profiles. Their densities are closely correlated to the organic matter, which is mainly concentrated in the most superficial layer of the forest soil, i.e. the litter.

The abundance of animals in the soil is negatively correlated with soil depth, since it depends on the availability of nutrient substances and oxygen. In forest soils, more than 60% of the ground fauna is typically concentrated in the first 10 cm of the superficial horizon (Peterson & Luxton, 1982). Soil arthropods make an enormous contribution to organic matter decomposition, breaking down and transforming all the material coming from the epigeal part of the forest. In fact, arthropods:

1. disintegrate the dead plant or animal tissues, transforming them into a substratum more easily attacked by micro-organisms,
2. selectively decompose and chemically modify part of the organic residues,
3. transform the plant residues into humic substances,
4. increase the surface area attacked by fungi and bacteria,
5. form complex aggregates of organic matter with mineral parts of the soil,
6. transport and mix the organic substances among the different layers,
7. disseminate propagules of fungi and stimulate their growth through symbiotic, commensal or phoretic relationships.

In temperate forests, the soil fauna remains active even in winter, under the snow cover and in the layers underlying the frozen litter. In these conditions, the breakdown and mineralization of organic matter is reduced to minimal levels.

Forest management practices can increase or reduce the ecological niches of the soil and thus influence the range of species. In environments less impacted by human activities, the diversity and stability of animal populations are correlated via interactions among the different functional groups, which tend to bring the system toward equilibrium and stability. Soil arthropods and nematodes are notoriously flexible in terms of their diet and it is not easy to assign a single trophic level to each group. According to the available resources, Acari Oribatida and Collembola can behave as detritivores but can also feed on fungi, algae and amorphous detritus. Among Acari Mesostigmata, some Uropodina species are detritivores, while others behave as specialized predators. Likewise some soil-living nematodes are phytophages, feeding on tree roots, while others are generalists and others still are preyed upon by Acari Parasitidae, Macrochelidae and fungi (Marinari Palmisano & Irdani, 1996). Acari, particularly the Oribatida (Figs. 11 and 12), are very important for the maintenance of soil productivity, actively participating in the breakdown of plant detritus, the vertical transport of organic matter and the formation of humus. Studies conducted in beech and pine woods of central Italy revealed the presence of 90 species belonging to 64 genera and 42 families (Nannelli, 1972, 1980): 72 species were recorded in a single temperate beech stand (Nannelli, 1990).

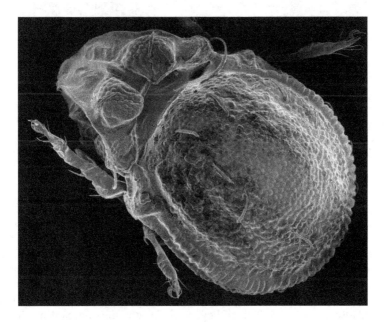

Fig. 11. A cryptostigmatic mite belonging to the genus *Carabodes*.

Fig. 12. A cryptostigmatic mite belonging to the genus *Pelops*.

Some soil mite groups have greater numbers of individuals in the boreal then in more southern zone. For exemple Oribatid and Prostigmatid mites are better represented in coniferous then in deciduous or tropical forest soil, and provide a significant share of total soil biomass (Petersen & Luxton, 1982).

Nematodes, perfectly adapted morphologically to interstitial habitats, are considered the most abundant animals on the Earth. The species found in soil are usually small, live in films of water lining soil and detritus particles, and feed on a wide range of substrata, including other nematodes preyed upon by many species of the same group. From the few data available, we can infer that the nematode communities of forest soils play a fundamental role in the control of microbial populations, carrying out this function even more effectively than protozoa. In some soils, microbivorous nematodes are the most numerous component of the nematode fauna (Anderson *et al.*, 1978) and act not only as consumers but also as vectors of both pathogenic and useful micro-organisms, in some cases showing specific associations with particular soil bacteria.

In table 3 are reported the distributions of nematode genera, in according of your feeding behaviour, in forest soils of central Europe.

Trophic group	Genera
Nematode bacteriophages	11-18
Nematode mycophages	2-5
Nematode phytophages	7-11
Nematode predators	2-8

Table 3. Distributions of nematode genera in groups feeding on different substrata recorded in forest soils of central Europe (from Wasilewska, 1979).

Large nematode populations are found in decomposing plant detritus, inside the tissues following the action of bacteria and protozoa. By their activity, they favour the humification and structural stability of the soils.

In general, micro-arthropods that feed on micro-organisms or behave functionally as detritivores are difficult to assign to only one of either primary or secondary decomposers. Both categories include Lumbricidae, Diplopoda, Isopoda, Collembola and Acari Oribatida, whereas Enchytraeidae and Coleoptera Elateridae are assigned to the secondary decomposers.

A similar model is used for soil-living predators, with several distinct levels: indeed, we can speak of first-, second- and third-level predators. Generally Chilopoda, Araneidae, Coleoptera Staphylinidae, Acari Gamasida and Uropodina, Pseudoscorpionida and Diplura are assigned to the first two levels, while third-level predators usually include higher animals such as micromammals and birds.

5. Conclusion

Animals included among consumers, demolishers and degraders actively participate in the cycles of organic matter and mineral elements, which without the intervention of these organisms would be destined to prolonged immobilization, especially in regard to the amounts conserved in durable wood structures (Dajoz, 2000). These organisms also establish complex interactions with predators and parasitoids forming part of the next trophic levels, which in turn are often subjected to predation and parasitism. This system of relationships is fundamental for the maintenance of homeostasis of forest ecosystems since it allows the species involved in the consumption and recycling of organic matter to play their role while at the same time maintaining their population numbers within strict limits so as not to jeopardize the equilibrium of the entire system. The last aspect is particularly important because of the necessity to keep primary consumer arthropods and nematodes, which feed on green photosynthesizing organs and subcortical tissues, within the carrying capacity of the ecosystem.

6. Acknowledgements

We thank our nematologist colleagues Laura Ambrogioni and Beatrice Carletti of the CRA-ABP of Florence and Teresa Vinciguerra of the University of Catania for providing us with material and photos.

7. References

Anderson, R.V.; Elliot, E.T.; Mc Clellan, J.F.; Coleman, D.C.; Cole C.V. & Hunt H.W. (1978).
    Trophie interactions in soils as they affect energy and nutritient dynamies. III.
    Biotic interactions of Bacteria, Amoebae and Nematodes. *Microbial Ecology*, Vol. 4,
    pp. 361-371

Axelsson, B.; Lohm, U.; Nilsson, A.; Persson, T. & Tenow, O. (1975). Energetics of a larval
    population of Operophthera spp. (Lepidoptera, Geometridae) in central Sweden
    during a fluctuation low. *ZOON J. Zool. Inst. Zool. Univ. Upps*, Vol. 3, No.1, pp. 71-
    84

Bachelier, G. (1971). La vie animale dans les sols. I. Determinisme de la faune des sols, In: *La
    vie dans les sols, aspects nouveaux études expérimentales*, P. Pesson, (Ed.), 3-43,
    Gautithier Vuillars Editeur, Paris

Bernardi, W.; Pagliano, G.; Santini, L.; Strumia, F.; Tongiorgi Tommasi, L. & Tongiorgi, P.
    (1997). Natura e immagine. Il manoscritto di Francesco Redi sugli insetti delle galle,
    Edizioni ETS, Pisa, 253 pp.

Binazzi, A. & Roversi, P.F. (1990). Notes on the biology and ecology of Diphyllaphis
    mordvilkoi (Aizenberg), the oak woolly aphid, in Central Italy. *Acta
    Phytopathologica et Entomologica Hungarica*, Vol. 25, No. 1-4, pp. 67-70

Cao, V. & Luciano, P. (2007). Heterocerous macrolepidopterans of Sardinia. I. Quercus
    pubescens phytocoenosis in the Gennargentu Mountains. *Redia*, Vol. LXXXIX
    (2006), pp. 23-33

Chararas, C. (1962). *Scolytides des conifères*. Lechevalier, Paris.

Cleveland, L.R., (1934). The wood feeding roach Cryptocercus, its Protozoa and the
    symbiosis between Protozoa and roach. *Mem. Amer. Ac. Arts Sc.*, Vol. 17, pp. 185-
    242

Covassi, M. & Poggesi, G. (1986). Notizie sulla presenza e l'ecologia di Elatophilus pini
    (Bär.) in Italia (Heteroptera, Anthocoridae). *Redia*, Vol. LXIX, pp. 1-10

Covassi, M.V. & Marinari Palmisano, A. (1997). Relazioni tra nematodi ed artropodi negli
    ecosistemi forestali. *Atti I Congresso SISEF "La ricerca italiana per le foreste e la
    selvicoltura"*. Legnaro (PD), 4-6 giugno 1997, pp. 95-100

Dajoz, R. (2000). *Insects and Forests. The role and diversity of insects in the forest environments*.
    Intercept Lavoisier Publishing, Londres, Paris, New York: 668 pp.

Francardi, V. & Pennacchio, F. (1996). Note sulla bioecologia di Monochamus
    galloprovincialis galloprovincialis (Olivier) in Toscana e Liguria (Coleoptera
    Cerambycidae). *Redia*, Vol. LXXIX, No. 2, pp. 153-169

Funke, W. & Sammer, G. (1980). Stammablauf und Stammanflug von Gliederfüssern in
    Laubwäldern (Arthropoda). *Entomol. Gen.*, Vol. 6, pp. 159-168

Gregori, E. & Miclaus, N. (1985). Studio di un faggeta dell'Appennino Pistoiese; sostanza
    organica del suolo e produzione di lettiera. *Annali Ist. Sper. Studio e Difesa Suolo*,
    Vol. XVI, pp. 105-118

Haugaasen, T. (2009). A Lepidopteran defoliator attack on Brazil nut trees (*Bertholletia excels*)
    in Central Amazonia, Brazil. *Biotropica*, vol. 41(3), pp. 275-278.

Kukor, J.J. & Martin, M.M. (1983). Acquisition of digestive enzymes by siricid woodwasps
    from their fungal symbiont. *Science*, Vol. 220, pp. 1161-1163

Lasker, R. & Giese, A.C. (1956). Cellulose digestion by the sylverfish Ctenolepisma lineata. *J. Exp. Biol.*, Vol. 33, pp. 542-553

Luciano, P. & Roversi, P.F. (2001). *Fillofagi delle querce in Italia*. Industria Grafica Poddighe, Sassari, 161 pp.

Marinari Palmisano, A. & Irdani T. (1996). Nematodi negli ecosistemi forestali. *Atti Giornate Fitopatologiche 1996*, Vol.1, pp. 253-260

Moran, V.C. & Southwood, T.R.E. (1982). The guild composition of arthropod communities in trees. *J. Anim. Ecol.*, Vol. 51, pp. 289-306

Nair, K.S.S. (2007). *Tropical forest insect pests: Ecology, impact, and management*. Cambridge University Press, Cambridge, 404 pp.

Nannelli, R. (1972). Ricerche sull'artropodofauna di lettiere forestali di pino e di quercia nei dintorni di Firenze. *Redia*, Vol. LIII, pp. 427-435

Nannelli, R. (1980). Composizione, successione e attività degli Oribatei (Acarina) nella lettiera di una pineta e di una querceta nei dintorni di Firenze. *Redia*, Vol. LXII, pp. 339-357

Nannelli, R. (1990). Studio di una faggeta dell'Appennino Pistoiese: composizione e successione dell'artropodofauna nella degradazione della lettiera. *Redia*, Vol. LXXIII, pp. 543-568

Nielsen, B.O. (1975). The species composition and community structure of the beech canopy fauna in Denmark. *Vidensk Meddel fra Dansk Natur Forening*, Vol. 138, pp. 137-170

Pennacchio, F.; Covassi, M.V.; Roversi, P.F.; Francardi, V. & Binazzi, A. (2006). Xylophagous insects of maritime pine stands attacked by Matsucoccus feytaudi Duc. In Liguria and Toscana (I) (Homoptera Margarodidae). *Redia*, Vol. LXXXVIII, pp. 1-7

Petersen, H. & Luxton, M. (1982). A comparative analysis of soil fauna population and their role in decomposition processes. *Oikos*, Vol. 49, pp. 287-388

Retallack, G.J. (1997). Early forest soils and their role in Devonian global change. *Nature*, Vol. 276, pp. 583-585

Rodin, L.E. & Bazilevich, N.T. (1967). *Production and mineral cycling in terrestrial vegetation*. Olivier and Boyd, Edinburg, 80 pp.

Roversi, P.F.; Covassi, M. & Toccafondi, P. (1990). Danni da Haematoloma dorsatum (Ahrens) su conifere (Homoptera Cercopidae). II. Indagine microscopica sulle vie di penetrazione degli stiletti boccali. *Redia*, Vol. LXXII, No. 2, pp. 595-608

Spies, T.A.; Franklin, J.F. & Thomas, T.D. (1988). Coarse woody debris in Douglas-fir forests of western Oregon and Washington. *Ecology*, Vol. 69, pp. 1689-1702

Stork, N.E. (1988). Insect diversity: facts, fiction and speculation. *Biol. J. Lin. Soc.*, Vol. 35, pp. 321-337

Thompson, V. (1994). Spittlebug indicators of nitrogen-fixing plants. *Ecol. Entomol.*, Vol. 19, pp. 391-398

Turner, B.D. (1975). Energy flow in arboreal epiphytic communities. An empiric model of net primary productivity in the alga Pleurococcus on Larch trees. *Oecologia*, Vol. 20, pp. 179-180

Wallwork, J.A. (1976). *The distribution and diversità of soil fauna,* Academic Press, London, New York, San Francisco, 355 pp.

Wasilewska, L. (1979). The structure and function of soil nematode communities in natural ecosystems and agrocenoses. *Polish Ecological Studies,* Vol. 5, No. 2, pp. 97-145

5

Vegetation Evolution in the Mountains of Cameroon During the Last 20,000 Years: Pollen Analysis of Lake Bambili Sediments

Chimène Assi-Kaudjhis
Université de Versailles Saint Quentin-en-Yvelines
France

1. Introduction

Tropical rainforests are the most biologically diverse ecosystems on the planet (Puig, 2001). In the highlands and mountains of western Cameroon (Central Atlantic Africa), a set of plateaus and mountains that contain large forested areas, this diversity is now subject to significant human pressure due to a large population engaged in agriculture and ranching. Momo Solefack (2009) shows that between 1978 and 2001 deforestation in Oku was 579 ha / year, with a annual rate of 4% increase. In addition to anthropogenic impacts, climate change plays a major role in influencing the distribution and composition of ecosystems (Walther et *al.*, 2002; Thomas et *al.*, 2004; Schröter et *al.*, 2005; Thuiller et *al.*, 2006).

The Bamenda Highlands have a particular forest characterized by the presence of one of the few African gymnosperms, *Podocarpus latifolius*. This species migrated from East to West Africa through Angola and then colonized the high mountains of Cameroon. It is currently restricted to altitudes above 1800 m, in the areas of Mount Oku and Kupé (Letouzey, 1968). Pollen data of Central Atlantic Africa have shown that this *Podocarpus* forest was once significantly more extensive than today during the last climatic cycle, especially during the last ice age (Dupont et *al.*, 2000; Elenga and Vincens, 1990; Maley and Livingstone, 1983). At this time, *Podocarpus* was present at low and mid altitudes mixed with the dense Guineo-Congolian forest. The expansion of *Podocarpus* into these areas did not end until very recently, about 3000 years ago (Vincens et *al.*, 2010). Such a distribution, involving the recent migration of the species to higher altitudes (White, 1993), suggests that such forests are likely refugia. This chapter, based on a sedimentary sequence of 14 m taken at Lake Bambili covering the last 20,000 years, presents the first palynological data from altitude in this region. Preliminary analysis of data from Bambili has been presented by Assi-Kaudjhis et *al.* (2008). The aim of this paper is to study the development and evolution of mountain forest in Cameroon over this interval.

2. Location, climate, and vegetation of bambili

Lake Bambili (05°56'11.9 N, 10°14'31.6 E, 2273 m asl) is a crater lake that lies in the volcanic zone of Cameroon (Figures 2 and 3) in the Bamenda Highlands and Bamboutos Mountains.

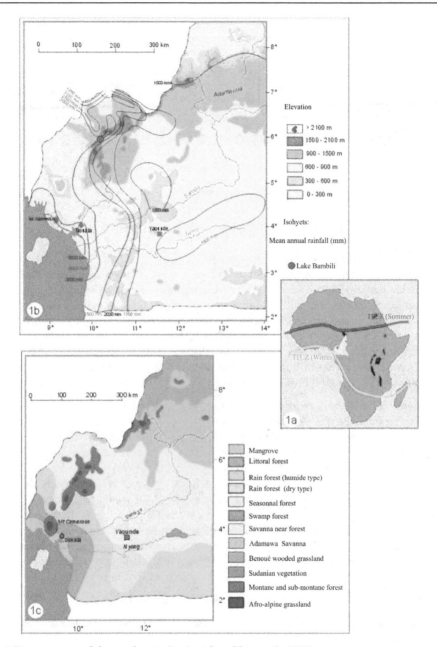

Fig. 1. Presentation of the study site (in Assi-kaudjhis et *al.*, 2008).
1a: Location of Lake Bambili, Cameroon; red line: ITCZ boreal summer; blue line: boreal winter ITCZ.
1b: Topography and rainfall in the study area.
1c: Phytogeography of the study area (Letouzey, 1968).

In northeastern Bambili, Mount Oku rises to 3011 m asl, the second highest peak in the country after Mount Cameroon (4070 m asl). The lake is part of a complex formed by two adjacent craters separated by about 45 m in altitude. The highest crater today is a swamp that discharges water into the crater below. The lower crater contains a lake of about 3 m depth. The lower crater contains a lake of about 3m depth with a single outlet to the northeast. The lake margin is narrow, consisting of a strip of herbaceous sub-aquatic vegetation, such as Cyperaceae and ferns, growing on peat-rich soil.

Fig. 2. The crater lake Bambili (05 ° 56'11.9 N, 10 ° 14'31.6 E, 2273 m asl). The red dot indicates the core location in the lower crater lake and the yellow dot is the second core location in the swamp of the second upper crater.

Centrally located in the Guineo-Congolian region, Cameroon has a relatively humid climate due to the locationof the country to the Gulf of Guinea (Suchel, 1988) which is responsible for the long rainy season over 4 / 5 of the country (western and southern regions) and slightly drier tropical climates in the north of the Adamawa plateaus. The seasonal alternation of southwesterly moisture flux and northerly dry winds, called the Alizé, creates a wet season from March to October and a more variable drier period during the rest of the year. The influence of altitude and distance from the coast result in lower precipitation in Bambili than Douala (2280 mm in Bamenda at 1370 m asl and 2107 mm in Bafoussam at 1411 m asl) and lower average temperatures (19°C to 20°C in Bamenda and Bafoussam against 26°C in Douala). Dry season precipitation from November to February is below 50 mm (Web LocClim, FAO, 2002). During the rainy season, precipitation is as high as 400 mm per month with temperatures fluctuating between 18.1 and 21.2°C.

Cameroon vegetation (Figures 3a and 3b) was described by Letouzey (1968, 1985) and White (1983). It is divided by latitude and altitude. At Bambili, aquatic vegetation grows in bands at the lake margin in the area of permanent open water: after *Nymphaea* sp. (Nymphaeaceae) on the edge of open water is a belt of Cyperaceae, ferns and aquatic plants, then, on the dry ground appear species of Poaceae.

Fig. 3a. Satellite Photo of crater lake Bambili (05°56'11 .9 N, 10°14'31 .6 E, 2273m asl).The dashed line connecting Bambili to Lake Oku is 36.5 km. Dark Red = forest; Light red = mosaic of fallow and degraded forest; Dark Blue = marshland; Black = open water; White = bare soil.

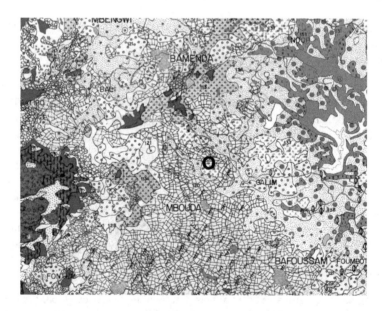

Fig. 3b. Vegetation map of the Bambili region (Letouzey, 1985) Green = moist forests; Light pink = fallow; Red = montane forest; Violet = submontane forest; Blue = marshland; Yellow = edge of the montane forest; Light yellow = savanna. Black circle indicates the location of Lake Bambili.

In general, the distribution of the plant communities of Cameroon at altitude is as follows:

Above 2400 m asl: afro-alpine
Alchemilla fisheri subsp. *camerunensis* (Rosaceae)
Agrostis mannii (Poaceae)
Veronica mannii (Asteraceae)...
From 2400 to 1600 m asl: montane forest
Podocarpus latifolius (Podocarpaceae)
Rapanea melanophloeos (Myrsinaceae)
Olea capensis (Oleaceae), Hypericum sp. (Ulmaceae)
Nuxia congestata (Loganiaceae)
Embelia sp. (Myrsinaceae), Celtis (Ulmaceae)
Clematis simensis (Ranunculaceae)
Hypericum sp. (Hypericaceae)
Syzygium staudtii (Myrtaceae)
Gnidia glauca (Thymelaeaceae)
Schefflera abyssinica, S. mannii (Araliaceae)
Arundinaria alpina (Poaceae)
From 1600 to 800 m asl: sub-montane forest (boundary between Guineo-Congolian and Afro-montane regions)
Olea hochstetteri (Oleaceae)
Schefflera abyssinica (Araliaceae)
Microglossia densiflora (Asteraceae)
Maytenus undata (Celastraceae)
Hypericum revolutum (Hypericaceae)
Prunus africana (Rosaceae)
Alchemilla fischeri (Rosaceae)...
Combined with elements of semi-deciduous rain forest
Polyscias fulva (Araliaceae)
Santiria trimera (Burseraceae)
Bridelia speciosa (Euphorbiaceae)
Uapaca sp. (Euphorbiaceae)
Leonardoxa africana (Caesalpiniaceae)
Celtis africana (Ulmanceae)
Anthocleista (Loganiaceae)...
Savannas and herbaceous layer
Annona senegalensis (Annonaceae)
Bridelia ferruginea (Euphorbiaceae)
Cussonia djalonensis (Araliaceae)
Terminalia avicennioides (Combretaceae)
Vernonia turbinata (Compositae)
Hymenocardia acida (Euphorbiaceae)

The organization of vegetation, based on Letouzey's studies (1968). But now, it is a theory, as environmental conditions have deteriorated (Table I). Thus, analysis of satellite images of

1998 and 2003 show that the forest cover around Bambili has deteriorated sharply in recent decades (Assi-Kaudjhis, 2011).

	1988	%	2003	%
Savanna, grassland and crops	3676.17	13.3	11615.65	42.03
Residential areas and bare soil	2196.79	7.95	4 766.18	17.25
Forest	3660.14	13.23	2472.65	8.95
Degraded forest	18086.1	65.40	8754.07	31.66
Lakes	31.81	0.12	29.31	0.11

Table I. Area (ha) of different units of land between 1988 and 2003.

3. Materials and methods

Two cores of 13.5 m and 14.01 m were taken a few meters apart at Lake Bambili in February 2007 and January 2010, respectively (Figure 4). The cores were taken using a Russian manual corer (Jowsey, 1966) in sections of 60 cm in length. The two sedimentary sequences were combined into a single sequence on the basis of benchmark levels identified in each, the depth (mcd) was calculated. The cores were sampled every 5 to 10 cm for pollen analysis.

Fig. 4. Coring on Lake Bambili using the Russian manual corer.

Overall, the sediments are composed of organic material and clay. From 0 to 635 cm, organic rich brown peat dominates becoming more compact and darker towards the base. Between 635 and 657 cm, sediments are mostly organic rich with centimeter-sized nodules of gray-green clay. Finally, 657 cm to the base of the sequence is an organic-rich compact black clay. A charcoal layer is observed at 1355 cm.

Laboratory number	Depth (cm) of the sample	Nature of the sample	Radiocarbon age BP			cal age-	cal age+	Average age (years cal BP)
SacA 8485	199-200	peats	1125	±	30	980	1037	1008.5
SacA 8486	299-300	peats	1745	±	30	1615	1704	1659.5
SacA 8487	399-400	peats	2170	±	30	2124	2301	2212.5
SacA 8488	499-500	peats	2315	±	30	2329	2352	2340.5
SacA 8489	599-600	peats	2485	±	30	2491	2707	2599
SacA 8490	699-700	peats	3175	±	30	3370	3442	3406
SacA 8491	799-800	peats	5515	±	30	6284	6318	6301
SacA 8492	899-900	peats	7255	±	35	8014	8156	8085
SacA 8493	999-1000	peats	8160	±	45	9015	9135	9075
SacA 8494	1099-1100	peats	10050	±	45	11404	11705	11554.5
SacA 8495	1198-1199	peats	11560	±	50	13323	13439	13381
SacA 10870	1248-1249	peats	12550	±	50	14605	14946	14775.5
SacA 10872	1348-1349	peats	14330	±	60	16950	17387	17168.5
Laboratory number	Depth (cm) of the sample	Nature of the sample	Radiocarbon age BP			cal age-	cal age+	Average age (years cal BP)
BB01-2010	1325	peats	14150	±	60	17032	17405	17218.5
BB01-2010	1355	charcoal	15020	±	60	18062	18505	18283.5
BB01-2010	1377	peats	15840	±	70	18840	19258	19049
BB01-2010	1410	peats	16910	±	70	19962	20256	20109

Table II. Radiocarbon dates and age model of the two cores taken in 2007 and 2010 at Bambili.

Seventeen AMS dates were performed, which show a continuous deposition for the last 20,000 years with a sedimentation rate ranging from 0.208 cm per year between 0 and 650 cm and 0.05 cm per year from 650 cm to the base (Table II). Radiocarbon measurements were calibrated using the CALIB software version 5.0. (Stuiver et al., 2005). The samples for pollen analysis were chemically treated with hydrochloric acid (HCl) and hydrofluoric acid (HF) according to the conventional method described by Faegri and Iverson (1975) preceded by a sieving at 250 microns to remove coarse particles. Treatment was terminated by filtration at 5 microns.

A total of 141 samples were analyzed with an average time resolution of 146 years. Counts ranged from 303 to 1500 pollen grains according to the richness of each sample, and 203 pollen taxa were identified. Data are presented as a diagram drawn on the basis of percentages calculated on a sum excluding aquatic plants and ferns.

The CONISS program used in the Tilia program (Grimm, 1987) was employed for the subdivision of the pollen diagram in zones. Analysis of palynological richness (Birks and Line, 1992), in order to estimate biodiversity, was performed with the software PSIMPOLL (http://www.chrono.qub.ac.uk/psimpoll/psimpoll.html).

4. Results

The microflora (Table III) consists of 119 taxa of trees, lianas and palms, together with 36 herbaceous taxa and 32 undifferentiated taxa, which correspond to plants that may be trees or herbs. Finally, 10 taxa correspond to aquatics plants, 4 to ferns and one to a plant parasite.

Family	Taxon	Habit	AA	FM	FSM	FSD	FDH	SAV	STP
LOGANIACEAE	Nuxia-type	A	X	X	X				
POLYGALACEAE	Polygala-type	AL	X	X	X			X	
ROSACEAE	Rubus	AL	X	X	X				
ROSACEAE	Rosaceae undiff.	I	X	X	X				
SOLANACEAE	Solanum-type	I	X	X	X				
URTICACEAE	Urticaceae undiff.	I	X	X	X				
BALSAMINACEAE	Impatiens	N	X	X	X	X			
RANUNCULACEAE	Thalictrum	N	X	X	X				
ERICACEAE	Ericaceae undiff.	A	X	X					
ERICACEAE	Erica-type	A	X	X					
MYRICACEAE	Myrica	A	X	X					
MYRSINACEAE	Rapanea	A	X	X					
RANUNCULACEAE	Ranunculaceae undiff.	A	X	X					
THYMELAEACEAE	Gnidia-type	A	X	X					
MYRSINACEAE	Maesa	AL	X	X				X	
GENTIANACEAE	Gentianaceae undiff.	I	X	X			X		
HYPERICACEAE	Hypericum	I	X	X					
ROSACEAE	Alchemilla	N	X	X					
ACANTHACEAE	Isoglossa	N	X	X					
CRASSULACEAE	Crassula	N	X	X					
DIPSACACEAE	Dipsacaceae undiff.	N	X	X					
LOBELIACEAE	Lobelia	N	X	X					
RUBIACEAE	Galium-type	N	X	X					
POLYGONACEAE	Rumex	N	X		X				
MYRSINACEAE	Myrsine-type africana	A	X						
ASTERACEAE	Artemisia	I	X						X
RUBIACEAE	Anthospermum	I	X						
LAMIACEAE	Leucas-type	N	X					X	
PRIMULACEAE	Anagallis	N	X					X	
EUPHORBIACEAE	Flueggea	A		X	X	X	X	X	
SALICACEAE	Salix	A		X					
ACANTHACEAE	Mellera-type	I		X					
ARALIACEAE	Araliaceae undiff.	A		X	X	X		X	
ARALIACEAE	Polyscias	A		X	X	X		X	

Family	Taxon	Habit	FM	FSM	FSD	FDH	SAV	STP
BURSERACEAE	Santiria-type	A	X	X	X	X		
COMBRETACEAE	Combretaceae/Melastomataceae undiff.	A	X	X	X	X	X	
HYPERICACEAE	Harungana	A	X	X	X		X	
MELIACEAE	Carapa-type procera	A	X	X	X	X		
MELIACEAE	Entandrophragma-type	A	X	X	X	X		
MELIACEAE	Meliaceae undiff	A	X	X	X	X	X	
MORACEAE	Ficus	A	X	X	X	X	X	
MORACEAE	Trilepisium-type madagascariensis	A	X	X	X	X		
MYRSINACEAE	Embelia-type	A	X	X	X	X		
MYRTACEAE	Syzygium-type	A	X	X	X	X	X	
RUBIACEAE	Pavetta	A	X	X	X			
RUBIACEAE	Psydrax-type	A	X	X	X			
RUTACEAE	Zanthoxylum-type	A	X	X	X	X	X	
SAPINDACEAE	Allophylus	A	X	X	X	X	X	
STERCULIACEAE	Dombeya-type	A	X	X	X			
ULMACEAE	Celtis	A	X	X	X	X	X	
ULMACEAE	Celtis/Trema	A	X	X	X		X	
ARALIACEAE	Schefflera	A	X	X				
HYPERICACEAE	Psorospermum	A	X	X			X	
OLEACEAE	Olea capensis	A	X	X				
RUBIACEAE	Keetia-type cornelia	A	X	X				
SAPOTACEAE	Sapotaceae undiff.	A	X	X			X	
SIMAROUBACEAE	Brucea	A	X	X				
OLEACEAE	Olea europaea-type	A	X	X				
DILLENIACEAE	Tetracera	AL	X	X	X	X	X	
APOCYNACEAE	Landolphia-type	AL	X	X		X		
ACANTHACEAE	Justicia-type	I	X	X	X			
APOCYNACEAE	Apocynaceae undiff.	I	X	X	X	X	X	X
FABACEAE	Dolichos-type	I	X	X	X			
URTICACEAE	Laportea-type	N	X	X	X			
ACANTHACEAE	Hypoestes-type	N	X	X	X		X	
AMARANTHACEAE	Achyranthes-type	N	X	X			X	
RUBIACEAE	Spermacoce-type	N	X	X				
BEGONIACEAE	Begonia	N andNL	X	X	X			
PALMAE	Phoenix	PA	X	X	X			
LORANTHACEAE	Tapinanthus-type	Par	X	X				
ANACARDIACEAE	Lannea-type	A	X		X		X	
CELASTRACEAE	Celastraceae undiff.	A	X		X	X	X	
IRVINGIACEAE	Irvingia-type	A	X		X	X		
RUTACEAE	Clausena anisata	A	X		X		X	
EUPHORBIACEAE	Phyllanthus-type	A	X		X		X	
AQUIFOLIACEAE	Ilex mitis	A	X				X	

Family	Taxon	Habit	FM	FSM	FSD	FDH	SAV	STP
CELASTRACEAE	Maytenus	A	X				X	
PODOCARPACEAE	Podocarpus	A	X					
RHAMNACEAE	Rhamnus-type	A	X					
ROSACEAE	Prunus	A	X					
FLACOURTIACEAE	Flacourtia	A	X					
CELASTRACEAE	Cassine	AL	X		X	X		
MIMOSACEAE	Acacia	AL	X				X	X
MENISPERMACEAE	Cissampelos-type	AL	X				X	
FABACEAE	Indigofera	I	X		X		X	
RANUNCULACEAE	Clematis-type	L	X		X			
FABACEAE	Eriosema-type	N	X		X	X	X	
AMARANTHACEAE	Celosia argentea-type	N	X				X	
FABACEAE	Lotus-type	N	X					
GENTIANACEAE	Sebaea	N	X			X		
GENTIANACEAE	Swertia abyssinica-type	N	X					
GESNERIACEAE	Streptocarpus	N	X					
EUPHORBIACEAE	Alchornea	A		X	X	X	X	
EUPHORBIACEAE	Antidesma-type	A		X	X	X	X	
EUPHORBIACEAE	Uapaca	A		X	X	X	X	
LOGANIACEAE	Anthocleista	A		X	X	X	X	
ANACARDIACEAE	Pseudospondias-type	A		X	X			
ANACARDIACEAE	Sorindeia-type	A		X	X		X	
EUPHORBIACEAE	Macaranga-type	A		X	X			
EUPHORBIACEAE	Margaritaria discoidea	A		X	X		X	
MELIANTHACEAE	Bersama	A		X	X		X	
MIMOSACEAE	Entada-type	A		X	X		X	
MORACEAE	Myrianthus-type	A		X	X		X	
RUBIACEAE	Ixora-type	A		X	X			
STERCULIACEAE	Sterculia-type	A		X	X		X	
EUPHORBIACEAE	Croton-type	A		X	X		X	
CAESALPINIACEAE	Leonardoxa-type africana	A		X		X		
EUPHORBIACEAE	Bridelia-type	A		X		X	X	
EUPHORBIACEAE	Erythrococca-type3	A		X			X	
FABACEAE	Baphia-type	AL		X	X	X		
RUBIACEAE	Tarenna-type	AL		X				
CAPPARIDACEAE	Capparidaceae undiff.	I		X	X		X	X
EUPHORBIACEAE	Acalypha	I		X	X			
STERCULIACEAE	Pterygota	N		X	X			
MENISPERMACEAE	Stephania-type abyssinica	NL		X				
OCHNACEAE	Campylospermum	A		X				
MALPIGHIACEAE	Acridocarpus	AL		X				

Family	Taxon	Habit	FSD	FDH	SAV	STP
EUPHORBIACEAE	Drypetes-type	A	X	X		
EUPHORBIACEAE	Tetrorchidium	A	X	X		
MELIACEAE	Khaya-type	A	X	X	X	
MIMOSACEAE	Pentaclethra macrophylla	A	X	X		
MORACEAE	Antiaris-type toxicaria	A	X	X	X	
MYRISTICACEAE	Pycnanthus	A	X	X		
OLACACEAE	Strombosia	A	X	X		
RUBIACEAE	Bertiera	A	X	X		
RUBIACEAE	Pausinystalia-type	A	X	X		
ANACARDIACEAE	Rhus-type	A	X		X	
ARALIACEAE	Cussonia	A	X		X	
BALANITACEAE	Balanites	A	X		X	
CONNARACEAE	Cnestis-type	A	X		X	
DICHAPETALACEAE	Tapura fischeri-type	A	X			
EUPHORBIACEAE	Mallotus-type	A	X			
HYMENO-CARDIACEAE	Hymenocardia	A	X		X	
OCHNACEAE	Lophira	A	X		X	
OLEACEAE	Schrebera	A	X			
RUBIACEAE	Morelia-type senegalensis	A	X		X	
RUTACEAE	Teclea-type	A	X			
SAPINDACEAE	Aphania-type senegalensis	A	X		X	
SAPINDACEAE	Blighia	A	X			
SAPINDACEAE	Lecaniodiscus/Aphania senegalensis	A	X			
ULMACEAE	Holoptelea grandis	A	X		X	
CAESALPINIACEAE	Cassia-type	I	X		X	
TILIACEAE	Triumfetta-type	I	X		X	
SAPINDACEAE	Sapindaceae undiff.	AL	X	X		
OCHNACEAE	Sauvagesia erecta	N	X			
ANISOPHYLLEACEAE	Anopyxis klaineana	A	X			
CAESALPINIACEAE	Crudia-type	A		X		
EUPHORBIACEAE	Klaineanthus gaboniae	A		X		
EUPHORBIACEAE	Thecacoris-type	A		X		
RUBIACEAE	Adenorandia-type kalbreyeri	A		X		
SAPINDACEAE	Aporrhiza	A		X		
OLACACEAE	Olacaceae undiff.	AL		X		
EUPHORBIACEAE	Cyathogyne	N		X		
BURSERACEAE	Commiphora	A			X	X
CAPPARIDACEAE	Crateva adansonii	A			X	X
MENISPERMACEAE	Cocculus	A			X	X
RUBIACEAE	Mitragyna-type inermis	A			X	X
CAPPARIDACEAE	Maerua-type	A			X	
CAPPARIDACEAE	Boscia-type	I			X	X
LAMIACEAE	Basilicum polystachyon/Hoslundia opposita	I			X	

Family	Taxon	Habit	SAV	STP
VERBENACEAE	Lippia-type	I	X	
AMARANTHACEAE/CHENOPODIACEAE	Amaranthaceae/Chenopodiaceae undiff.	N	X	X
RUBIACEAE	Mitracarpus villosus	N	X	X
STERCULIACEAE	Hermannia-type	N	X	X
LAMIACEAE	Leonotis-type	N	X	
AMARANTHACEAE	Aerva-type	N		X
PROTEACEAE	Faurea-type	A	X	
VITACEAE	Cissus	I	X	
COCHLOSPERMACEAE	Cochlospermum	I	X	X
SAPINDACEAE	Pappea capensis	A	X	
CAESALPINIACEAE	Parkinsonia aculeata-type	A	X	
ANACARDIACEAE	Anacardiaceae undiff.	A		
ACANTHACEAE	Acanthaceae undiff	I		
APIACEAE	Apiaceae undiff.	I		
ASTERACEAE	Asteraceae undiff.	I		
ASTERACEAE	Cichoriae undiff.	I		
EUPHORBIACEAE	Euphorbiaceae undiff.	I		
EUPHORBIACEAE	Euphorbia-type	I		
FABACEAE	Fabaceae undiff.	I		
LAMIACEAE	Lamiaceae undiff.	I		
RUBIACEAE	Rubiaceae undiff.	I		
SOLANACEAE	Solanaceae undiff.	I		
MONOCOTYLEDONAE	Monocotyledones	I		
BRASSICACEAE	Brassicaceae undiff.	N		
ASTERACEAE	Centaurea-type	N		
PLANTAGINACEAE	Plantago	N		
POACEAE	Poaceae undiff.	N		
EUPHORBIACEAE	Ricinus communis	N		
CYPERACEAE	Cyperaceae undiff.	Nq		
HALORRHAGACEAE	Laurembergia tetrandra	Nq		
HYDROCHARITACEAE	Ottelia-type	Nq		
NYMPHAEACEAE	Nymphaea	Nq		
ONAGRACEAE	Ludwigia-type	Nq		
ONAGRACEAE	Onagraceae undiff.	Nq		
POLYGONACEAE	Polygonum senegalense-type	Nq		
POTAMOGETONACEAE	Potamogeton	Nq		
TYPHACEAE	Typha	Nq		
XYRIDACEAE	Xyris	Nq		
Monoletes smoothferns		Sp		
Monoletes ferns NL		Sp		
Triletes smooth ferns		Sp		
Triletes ferns NL		Sp		

Table III. List of pollen taxa determined at Bambili. Taxa are ranked according to membership of the corresponding plants to specific vegetation types: AA: Afro-montane;

FM: montane forest; FSM: sub-montane forest; SDF: semi-deciduous forest; FDH: rainforest ; SAV: savannah; STP: steppe. The plant habit is based on definitions of Vincens et *al.* (2007) A: trees; AL: trees and / or lianas; PA: palm; Par: parasites; N: herbs; NL: herbaceous lianas; Nq: aquatic herbs; I: undifferentiated; Sp: fern.

The pollen diagram shows changes in tree taxa percentages between 10.56% (16,593 cal yrs BP) and 95.28% (7032 cal yrs BP) (Figure 4). A number of these taxa belong to the montane forests and mid-altitude dense forest, which are present in almost all samples analyzed. These include *Podocarpus, Schefflera, Alchornea, Celtis, Embelia, Maesa, Macaranga*-type, *Olea capensis, Ficus, Syzygium* and *Rapanea*. However, these percentages are highly variable: from 0.2 to 45.75%. Based on these changes as well as that of the appearance and disappearance of other characteristic taxa (*Aerva, Alchemilla, Artemisia, Farsetia, Hypericum, Ilex mitis, Gnidia*-type and *Myrica*), five pollen zones were distinguished for the sequence:

Zone I: 14.01 to 13.21 mcd (20,109-17,192 cal yrs BP)

This zone is characterized by high percentages of herbaceous plants including Poaceae undiff., which decreases from the base (75.68%) to the top (34.44%), and Asteraceae undiff. (30.29%). The tree percentages do not exceed 26.64%.

This zone also includes many taxa of trees and herbaceous plants that characterize today's open spaces, such as savannas and steppes, including *Lannea* (1.22%), *Commiphora* (0.44%) Capparidaceae undiff. (0.34%) and *Aerva*-type (6.15%). The afro-alpine meadow and tree line are also represented by a number of taxa: *Myrica* (16.99%), Ericaceae undiff. (3.09%), *Artemisia* (1.68%), *Maesa* (1.29%), *Gnidia*-type (0.93%) linked to montane elements, *Nuxia* (9.31%) *Rapanea* (1.25%), *Rubus* (2.37%), and *Carapa procera* (1.82%).The trees of the sub-montane forest and semi-deciduous forest are present in proportions not exceeding 2.5%. These are primarily *Antiaris, Antidesma, Lophira, Pausynistalia, Trilepisium madagascariensis*-type, and Sapindaceae undiff.

Based on the variations of the main taxa of this zone (Asteraceae, *Aerva*-type, *Anthrospermum, Myrica*), two sub-areas were identified:

Subzone Ia: 14.01 to 13.63 mcd (20,109-18,589 cal yrs BP)

This subzone is characterized by the maximum percentages of Asteraceae undiff. (30.29%) at 13.495 mcd (19,049 cal yrs BP). The vegetation is dominated by herbaceous taxa and shrubs of the open savanna or forest edges: Amaranthaceae / Chenopodiaceae undiff. (2.30%), Asteraceae undiff. (30.29%), *Aerva*-type (2.63%), Solanaceae undiff. (3.13%), Lamiaceae undiff. (4.15%), *Achyranthes*-type (0.95%), *Lannea* (0.41%) *Boscia* (0.41%), *Crudia* (0.35%) and Urticaceae undiff. (0.74%).

Subzone Ib: 13.63 to 13.21 mcd (18,589-17,192 cal yrs BP)

All trees decreased from 32.89% to 12.84%, while herbaceous taxa increase from 44.40% to 80.41%. This sub-zone is characterized by the successive peaks, between 18,436 and 18,167 cal yrs BP, of herbaceous taxa of the subalpine vegetation type: *Anthospermum* (6.95%), followed at 17,640 cal BP by a first peak of *Olea capensis* (4%) and *Schefflera* (6.22%).

Zone II: 13.21 to 12.295 mcd (17,192-14,320 cal yrs BP)

This area is characterized by high values of Poaceae undiff. (68.58%) associated with *Aerva* at maximum values (6.15%). Asteraceae remains significant throughout the zone.

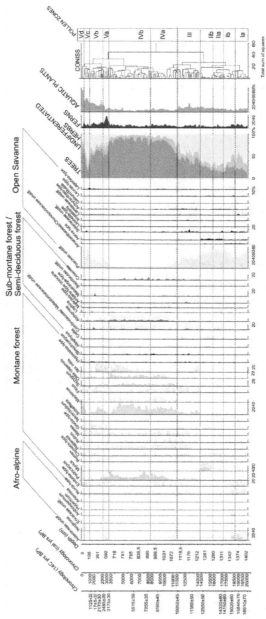

Fig. 4. Pollen diagram of Lake Bambili (Cameroon). This diagram shows, from left to right, radiocarbon ages, calibrated ages, depths, and percentages of major taxa presented in terms of large groups of regional vegetation calculated on a sum excluding aquatic plants and ferns. Poorly represented taxa are exaggerated by a factor of 10 and superimposed on the initial values (in dark). At the far right is a diagram of variations of AP / NAP. Pollen zones were delineated by applying the CONISS software (GRIMM, 1987).

Vegetation Evolution in the Mountains of Cameroon During the Last 20,000 Years:
Pollen Analysis of Lake Bambili Sediments

123

Other savanna and steppe taxa are also present: Amaranthaceae / Chenopodiaceae undiff (1.39%), *Commiphora* (0.48%), *Mitragyna inermis*-type (0.24%), *Faurea* (0.45%) *Crateva adansonii* (0.23%). Progressions of *Myrica*-type and *Aerva* determine two sub-zones:

Sub-zone IIa: 13.21 to 12.83 mcd (17,192-15,916 cal yrs BP)

In this sub-zone, the montane forest trees and afro-alpine meadow values remain small or decline sharply from the previous zone: *Podocarpus* (3.56%), *Myrica* (2.85), *Schefflera* (2.49%), *Syzygium* (1.85%), *Olea capensis* (1.43%), *Embelia*-type (2.31%) and Ericaceae undiff. (1.1%) while only *Anthospermum* disappears.

Sub-zone IIb: 12.83 to 12.295 mcd (15,916-14,320 cal yrs BP)

This sub-zone is distinguished by very high values of *Myrica* between 15,496 and 14,320 cal yrs BP (16.99%). All trees increase (31.84%); however, this primarily montane taxa (30.94%; *Carapa* type *procea, Clausena anisata, Clematis, Embelia, Ilex mitis,* and *Lannea*), and sub-montane taxa (7.82%; *Macaranga, Bredelia*-type, *Croton*-type, *Erythrococca*). Most savanna and steppe components disappear, except for *Mitragyna inermis*-type, *Cohlospermum, Commiphora,* and *Crateva adansonii,* with percentages not exceeding 1%.

Zone III: 12.295 to 11.005 mcd (14,320-11,572 cal yrs BP)

This zone is marked by the drastic reduction of Poaceae undiff. percentages by more than 11.13% (minimum at 11,757 cal yrsq BP). The percentages of all trees increase, especially *Schefflera* (19.07%) and *Podocarpus* (7.45%). This zone is also characterized by the final disappearance of *Myrica*, which is replaced by *Ilex mitis* (5.11%).

Between 14,200 and 13,602 cal yrs BP, tree percentages decreased significantly, especially *Schefflera* and *Podocarpus* which represent only 12.03% and 7.45%, respectively.

Zone IV: 11.005 to 6.80 mcd (11,572-3,252 cal yrs BP)

The increase in trees initiated in the previous zone continues in zone IV with percentages reaching the maximum value of 92%. Conversely, Poaceae undiff. disappears along with all elements of non-native steppe and savanna. The changing environment consists of two phases:

Subzone IVa: 11.005 to 9.205 mcd (11,572-8,252 cal yrs BP)

The disappearance of *Ilex mitis* precedes the maximum phase of forest expansion. The general increase of trees results in the appearance and / or expansion of a series of montane forest taxa: *Olea capensis* (15.61%), *Syzygium* (12.62%), *Maytenus* (4.36%), *Ficus* (6%), *Embelia* (3.40%), *Rubus* (5.38%), *Achyranthes*-type (1.38%), *Clematis* (1.84%), *Celtis* (4.34%) *Podocarpus* (32.13%) and *Schefflera* (36%). The sub-montane forest taxa are also increasing: *Alchornea* (7.36%), *Mallotus*-type (3.35%), *Macaranga*-type (3.88 %), *Erythrococca*-type (2.07%), and *Streptocarpus* (4.15%).

Subzone IVb: 9.205 to 6.80 mcd (8,292-3,252 cal yrs BP)

This sub-zone is marked by the re-emergence of *Ilex mitis* (1.74% to 10.15%) and the irregular representation of different forest taxa, all trees ranging from 86.93% to 75.80%. Montane forest taxa, such as Ericaceae undiff. (1.32%), *Isoglossa* (0.23%), *Gnidia* (1.15%), *Maesa* (4.36%), *Rapanea* (3.48%) and *Nuxia*-type (1.75%), have relatively high percentages.

Zone V: 6.80 to 0 mcd (3,252 cal yrs BP-Present)

The pollen spectra of this area reveal a severely depleted forest with percentages of trees that represent only 39% on average. Many taxa completely disappear (*Indigofera, Blighia,* Sapindaceae undiff., *Brucea, Sauvagesia, Faurea, Salix*). Others remain present in significant percentages such as *Podocarpus* (up to 19.43%), *Olea capensis* (12.24%), *Schefflera* (45.25%), *Ilex mitis* (13.27%), Asteraceae undiff. (15.74%), and Ericaceae undiff. (4.45%). In addition, *Achyranthes* (1.53%) and *Laportea*-type (6.33%), both characteristic of degraded environments, along with *Gnidia* (6.08%), *Impatiens* (1.52%) and *Ricinus* (0, 60%) occur and progress in this area. The different variations of the percentages of major taxa suggest four sub-zones:

Subzone Va: 6.80 to 5.805 mcd (3,252-2,549 cal yrs BP)

There is a change in vegetation in this sub-zone marked by the loss of forest as percentages fall to 46% (*Alchornea* (1%), *Allophylus* (0.29%), *Antiaris* (0.30%), *Antidesma*-type (0.81%), *Carapa*-type *procea* (0.34%), *Celtis* (0.40%), *Clausena anisata* (0.67%), Combretaceae undiff. (0.30%), *Cussonia* (0.64%), *Drypetes* (0.29%), *Embelia* (0.30%), Ericaeae undiff. (0.39%), *Erythrococca*-type (0.34%), *Ficus* (0.30%), *Gnidia* (0.31%), *Podocarpus* (5.43%), *Olea capensis* (0.30%), and *Schefflera* (7.66 %)). Poaceae undiff., Asteraceae undiff., and ferns increase dramatically to 36.50%, 5.53% and 15.35%, respectively.

Subzone Vb: 5.805 to 1.905 mcd (2,549 -960 cal yrs BP)

Forest vegetation recovers but is impoverished and deteriorates very quickly at 960 cal yrs BP. This sub-zone also shows the importance of indicators for open environments: *Celtis* (11.84%), *Laportea* (6.33%), Solanaceae undiff. (2.59%), Poaceae undiff (48.95%), *Macaranga* (10.10%), Lamiaceae undiff. (2.17%), *Hypoestes*-type (2.55%), and *Clematis* (2.36%). The majority of the montane taxa cited in the previous sub-zone regress: *Podocarpus* (0.51%). *Maytenus* (0.28%), *Embelia*-type (0.42%), *Syzygium* (1.27%), *Schefflera* (0.84%), *Ilex mitis* (1.68%), *Antiaris* (0.76%), *Rapanea* (0.38%) and *Cussonia* (0.31%).

Subzone Vc: 1.905 to 0.605 mcd (960-266 cal yrs BP)

This subzone marks the minimum representation of trees in the Holocene with a percentage of 18%. Herbaceous plants increase, especially Poaceae undiff. (78%), Asteraceae undiff. (4.11%), Urticaceae (*Laportea*-type) (1%) and Solanaceae undiff. (5.80%). Forest elements such as *Olea capensis* (1.61%), *Podocarpus* (1.36%), *Maytenus* (0.4%), *Syzygium* (1.27%), and *Ficus* (0.34%) as well as afro-alpine elements such as Ericaceae undiff. (0.34%), *Hypericum* (0.40%), and *Isoglossa* (0.80%), are present in very small quantities.

Subzone Vd: 0.605 to 0 mcd (266 cal yrs BP-Present)

The top of this record is differentiated by the extreme poverty of the forest flora dominated by *Schefflera* (45%) and *Embellia* (11%), associated with *Syzygium* (7.07%), *Hypoestes*-type (4.04%), *Clematis* (2.26%), *Celtis* (3.80%), *Ilex mitis* (0.45%), *Cussonia* (3.61%), *Achornea* (4.22%), and *Macaranga*-type (1.1%). Poaceae undiff. remain at a high level and Solanaceae undiff. increases (5.5%).

Vegetation Evolution in the Mountains of Cameroon During the Last 20,000 Years:
Pollen Analysis of Lake Bambili Sediments

125

5. Discussion: The evolution of vegetation cover

5.1 An open and degraded vegetation during the ice age

Forest cover

At Bambili, forest cover was extremely low during the late glacial period (18,000-23,000 cal yrs BP). The vegetation is dominated by herbaceous plants, as is the case in all West African sites regardless of altitude: the swamp Shum Laka (Kadomura, 1994) at 1200 m asl on the Bamenda plateau at Barombi Mbo at the foot of Mount Cameroon (Maley and Brenac, 1998) and Lake Bosumtwi, Ghana (Maley, 1987) located in low and mid altitudes. The pollen studies from the mountains of East Africa also reflect the general degradation of tropical forests during this period (Livingstone, 1967, Coetzee 1967, Hamilton, 1982). Three pollen sites located in southern hemisphere, however, show a different pattern. At Ngamakala, a forest sites in the southern Congo located 400 m asl, the environment, although degraded, remains forest during the glacial period until 13,000 yrs BP (Elenga et al., 1994). At Kisiga Rugaro, a forested part of the Eastern Arc Mountains shows a certain environmental stability during the whole glacial period (Mumbai et al. 2008). They note, however, that the herbaceous plants recorded their highest percentages between 19,000 and 14,000 cal yrs BP , which could correspond to drier and colder conditions than even the preceding LGM. Mumbai et al. (2008) suggest that the relative stability of ecosystems during the last ice age is due to the influence of the Indian Ocean that would have allowed the maintenance of a rain forest while the regional climate of East Africa was dry. At Lake Masoko (9°20'S, 33°45'E, 840 m asl), Vincens et al. (2006) also identify a uninterrupted development phase of semi-deciduous forest from 23,000 - 11,800 cal yrs BP.

The composition of local vegetation

The composition of pollen spectra of the Last Glacial Maximum (LGM) at Bambili shows a mixture of floristic elements representative of distinct stages: the sub-montane forest, the semi-deciduous forest (*Antiaris, Antidesma, Lophira, Pausynistalia, Trilepisium madagascariensis*-type, Sapindaceae undiff.), the montane forest (*Nuxia, Rapanea, Rubus, Carapa procea, Myrica, Gnidia*-type, *Maesa*, Ericaceae undiff.) and Afro-alpine (*Artemisia, Alchemilla, Anthospermum, Hypericum, Rumex, Isoglossa*). Herbaceous taxa dominate open areas or forest edges. Among these are: Poaceae undiff., Cyperaceae undiff., Amaranthaceae / Chenopodiaceae undiff., Asteraceae undiff., Lamiaceae undiff., *Achyranthes*-type and Urticaceae undiff. Among the arboreal taxa, montane forest taxa dominated; however, their percentages are very low.

Distant contributions

The LGM is characterized by the presence of savanna and steppe taxa such as *Aerva*-type, *Boscia*-type, Capparidaceae undiff., *Commiphora, Crataeva adansonii, Crudia*-type, *Lannea*-type and *Maerua*-type that show the importance of long-distance aeolian input, related to an increased flow of trade winds (Sarnthein et al., 1981) and possibly the extension of Sudan-Zambezian vegetation zones due to dry conditions. A layer of charcoal was observed at 1355 cm. Dating of this layer yielded an age of 18 283 cal yrs BP (15 020 ± 60 ^{14}C BP) suggesting the importance of fire at this time in the environment near Lake Bambili.

5.2 The post-glacial forest colonization

Chronology of colonization forest

In the global post-glacial context, recolonization of forest is observed. At Bambili, this occurs in three distinct stages (Figure 5) interrupted by two phases of regression corresponding to the Heinrich event 1 (H1) and the Younger Dryas.

The first phase of colonization starts at 18,400 cal yrs BP with the appearance of *Anthospermum*. This taxon is noted by Livingstone (1967) as an important element of the first stage of colonization of lava fields in the region of the Virunga volcanoes in East Africa. In Cameroon, *Anthospermum camerounensis*, is a dwarf grass found in Afro-alpine vegetation types (Letouzey, 1968). This taxon is followed by a large increase of *Olea capensis* and *Schefflera*, forming a first phase of increase in trees. The increase of trees stops at 17 100 cal yrs BP, with the increase of *Aerva*-type, which indicates the strengthening of the boreal winter northeasterly wind flow from the Sahelian steppes and large fires are indicated by charcoal layers. This increase is contemporaneous with H1, the characteristics of climate on a global scale are drought (Mix et al., 2001; Kageyama et al., 2005; Timmermann and Menvielle, 2009).

Vegetation at Bambili remains relatively stable for about 1900 years when a second stage of forest colonization begins at 14,900 cal yrs BP. This stage is initiated by the appearance of *Myrica*, a fire-tolerant, sun-loving shrub common in clearings of the upper montane forests (Livingstone, 1967). This taxon is followed by another sun-loving species, *Ilex mitis*, and finally by *Olea capensis, Podocarpus* and *Schefflera*. This phase culminates in a very short period between 13,800 and 13,700 cal yrs BP, corresponding to the beginning of the warm Alleröd in high latitudes (Roberts et al., 1993, 2010). This increase of forest cover at Bambili is coeval with the start of the African Humid Period dated to 15,500 cal yrs BP by DeMenocal et al. (2000). This period was marked by increased flow of the Niger River (Lézine and Cazet, 2005) and the general rise in lake levels (Gasse, 2000; Shanahan et al., 2006). At Bambili, Stager and Anfang-Sutter (1999) found a positive P / E during this period. The presence of steppe taxa such as *Aerva* up to 11,700 cal yrs BP, however, shows that trade wind flow was still significant over this period. This is in contrast with the termination of Saharan dust transport to the ocean further north noted by DeMenocal et al. (2000).

Between 13,000 and 11,700 cal yrs BP, a second phase of forest decline is recorded. This phase does not correspond to the total destruction of the forest. It is characterized instead by a massive regression of *Schefflera*. All trees recorded a large decrease with the exception of *Podocarpus, Rapanea* and *Gnidia*-type. The permanence of steppe elements reflects the importance of continued northeasterly circulation. This phase corresponds to the YD (Younger Dryas) in the high latitudes (Roberts et al., 1993). As mentioned earlier, the YD is generally dry in tropical North Africa, reflected in the general lowering of lake levels (Gasse, 2000). This episode is marked by increased dust transport to the Atlantic recorded from the equator (Lézine et al., 1994) to the Saharan latitudes (DeMenocal et al., 2000). The YD signature on forest vegetation is however not very visible in the Ivory Coast as outlined Lezine and Le Thomas (1995). Despite evidence of increased sedimentological transport by the trade winds, the authors noted no major changes in the forest environment during this episode. The YD seems to be more clearly recorded in the vegetation surrounding lakes Barombi Mbo and Bosumtwi. Data from Barombi Mbo show an increase of spores (2 to over

15%), Poaceae (13 to 17%) and pioneers taxa such as *Trema, Macaranga, Mallotus* and *Alchornea* around 12,500 yrs BP (Maley and Brenac, 1998). At the same site, Lebamba et *al.* (2010) show, after ca 14,000 cal yrs BP, the decline of tropical rain forests and seasonal increases in savanna biomes. At Bosumtwi, there is an increase in grasses and sedges, while the tree taxa regress (Maley, 1991). From 11,500 cal yrs BP, *Schefflera* leads a new phase of forest expansion with a forest optimum dated between 10,000 and 8400 cal yrs BP. The development of the forest is gradual between 11,500 and 10,200 cal yrs BP and results in the expansion of montane taxa. Next, *Rubus, Rapanea, Embelia*-type and *Syzygium* appear in turn. The elements of the montane forest, in particular, *Schefflera, Podocarpus, Olea* and *Syzygium* rose steadily over this period. Submontane forest is also present through a number of taxa that appear sporadically, such as *Cussonia, Macaranga*-type, *Antiaris*-type *toxicaria*, and *Allophyllus*.

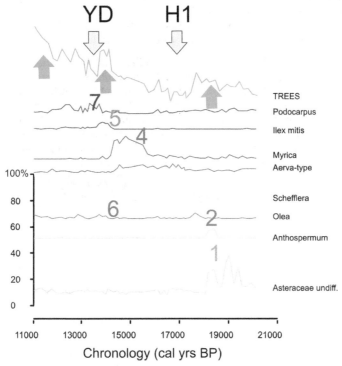

Fig. 5. Colonization of the postglacial forest. The curves are made from percentages of each taxon from 21,000 to 11,000 cal yrs BP. This colonization takes place in three phases forest interrupted by two regressions corresponding to H1 and the YD. It starts at 18,400 cal yrs BP with the appearance of *Anthospermum* (1) followed by *Olea capensis* (2) and *Schefflera* (3). The interruption to 17,100 cal yrs BP, marked by the increase of *Aerva*-type is correlated with H1. A second phase of forest colonization starts at 14,900 cal yrs BP with the appearance of *Myrica* (4) monitoring and *Ilex mitis Schefflera* (5), *Olea capensis* (6) and *Podocarpus* (7). Between 13,000 and 11,700 cal yrs BP, a second phase of regression is observed with decreasing forest trees. Then, from 11,500 cal yrsnBP, *Schefflera* (8) opens the Holocene forest expansion phase.

Example of two major forest taxa: *Olea* and *Podocarpus*

Comparison of pollen data from sites Barombi Mbo, Tilla and Bambili shows that, since 20,000 years BP, some plants have moved from low to high altitudes. This is the case of *Olea* (Figure 6) which shows percentages up to 35%, during the glacial period at Barombi Mbo (Maley and Brenac, 1998), indicating the presence of the plant source at low altitude near the lake. Its percentage decreased drastically from 12,000 cal yrs BP. At this time, *Olea* appears at Tilla, 700 m north of the volcanic line in Cameroon (Salzmann et *al.*, 2000) where it develops until it reaches 14%. It then extends to 2200 m asl at Bambili after 10,800 cal yrs BP where it remained until the end of the Holocene forest. This tree is still present in the mountains at the edge of montane forest and sub-montane, near the village of Oku. *Olea* is known to be a pioneer taxon, which explains its presence at the beginning of the forest recovery at Tilla, before the development of Euphorbiaceae (*Uapaca*) that characterize the forest in the Holocene. The behavior of *Olea* during the glacial-interglacial transition could be related to temperature changes. With rising temperatures after 12,000 years at low altitude, *Olea* migrates toward higher elevations to it current location.

Unlike the previous taxon, *Podocarpus* shows a similar pattern at both sites (Figure 7). It is not represented in the LGM sediments, and pollen percentages that reached Barombi Mbo are so low that it precludes the possibility of its presence at low altitude. At Bambili, it also reduced at this time. *Podocarpus* at Bambili increases from 10,000 cal yrs BP, then develops during two periods centered around 7560 cal yrs BP and 3360 cal yrs BP. It is interesting to note that these two peaks are also found at Barombi Mbo; however, here values are a full order of magnitude lower than at Bambili. This could support the hypothesis of two distinct phases of expansion of *Podocarpus* in altitude during the Holocene.

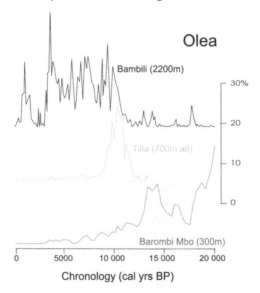

Fig. 6. The evolutionf *Olea*-type at Bambili, Tilla and Barombi Mbo. During the Last Glacial Maximum, *Olea*-type has a significant presence in Barombi Mbo until 12,000 cal yrs BP. The presence is then recorded at Tilla between 12,000 and 11,800 cal yrs BP and at Bambili at 10,800 to 3300 cal yrs BP.

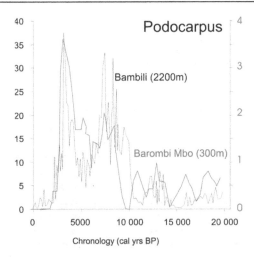

Fig. 7. The evolution of *Podocarpus* at Bambili and Barombi Mbo. The trend in *Podocarpus* expansion is similar between sites; however, the values are often an order of magnitude higher at Bambili.

The expansion of *Podocarpus* during the Holocene is also attested throughout Central Africa from the Atlantic site at 8 m asl at Ossa (Reynaud-Farrera, 1995; Reynaud-Farrera et *al.*, 1996), as well as those of Mboandong at 130 m asl (Richards, 1986), Mbalang at 1100 m asl (Vincens et *al.*, 2009), Njupi at 1108 m asl (Zogning et *al.*, 1997), Bafounda at 1310 m asl (Tamura, 1990), and Shum Laka (W-10) at 1355 m asl (Kadomura, 1994).

5.3 The optimum forest of the early Holocene

The optimum forest at Bambili occurs between 10,000 and 8400 cal yrs BP and is characterized by the dominance of tree taxa at values from 80-92%. Of the 118 total tree taxa determined to Bambili, this period includes more than half (67 taxa), the majority belonging to the montane and semi-deciduous forests. These are mainly *Schefflera, Syzygium, Podocarpus, Olea capensis, Rapanea* for the montane forest and *Macaranga, Celtis, Cussonia, Antiaris* for semi-deciduous forest. Elements of the open areas have disappeared (*Leonotis, Leucas, Boscia, Aerva*) or decreased (*Impatiens, Achyrantes*, Solanaceae undiff, *Galium / Rubia*).

In tropical Africa during the early Holocene, the percentages of pollen from woody genera are high throughout the region, indicating the expansion of forests into higher latitudes and altitudes (Lézine, 2007). In West Africa, the marine records suggest that the Guineo-congolian forests were not separated by the savanna corridor that exists today in Togo and Benin, the "Dahomey Gap" (Dupont et *al* ., 2000), which is confirmed by the analysis of lake sediments from Sele (Salzmann et *al.*, 2000). Tropical plants migrated northwards along the rivers and lakes stretching across the Sahel and the Sahara (Watrin et *al.*, 2009). The mangrove taxon *Rhizophora* also occupied many coastal areas northward to around 21°N (Lézine, 1997).

Bonnefille et *al.*, (1995) noted also a reduction in forest cover during the LGM and a discontinuous forest colonization at Rusaka Swamp (3°26'S, 29°37'E, 2070 m asl). Many authors link the expansion of Guineo-congolian forest and mangrove with a reinforcement of the Atlantic monsoon at the beginning of the Holocene (Marzin et Braconnot, 2009). Rainfall was significantly higher than today and seasonality was reduced as shown by the high lake levels (Gasse, 2000; Shanahan et al., 2006) and increased fluvial transport (Lézine and Cazet, 2005).

5.4 The destabilization of the forest in the mid-Holocene: The 8.2 event

The forest phase at Bambili is marked by a very short but indicated episode of regression which occurs at the time of the 8.2 event in the high latitudes (Von Grafenstein et al., 1998). It is reflected in the physiognomy of the forest by lower percentages of trees and the more or less pronounced decrease in values of some taxa. The most remarkable decreases are those of *Schefflera, Podocarpus, Rapanea,* and to a lesser extent those of *Olea capensis, Maesa, Syzygium, Ilex mitis, Nuxia,* and *Embelia,* all standard elements of the montane forest as well as those of *Cussonia, Alchornea,* and the submontane forest. This degradation could be caused by dry conditions as suggested by the lower lake levels at Lake Bosumtwi (Shanahan et al., 2006) related to the slowdown of the thermohaline circulation in the North Atlantic (Pissart, 2002).

5.5 The end of the Holocene forest at Bambili

After 8400 cal yrs BP, changes in taxa suggest some forest instability. This phase of disruption occurs gradually, leading to the brutal destruction of the forest at 3300 cal yrs BP. At this period, *Podocarpus* opposes *Schefflera, Syzygium* and *Alchornea.* Forest degradation begins at 4500 cal yrs BP, with the decline of *Schefflera* followed by that of *Olea capensis* at 3500 cal yrs BP and *Podocarpus* at 3300 cal yrs BP. The drastic reduction of montane forest elements at 3300 cal yrs BP probably favored soil erosion and sediment supply from the crater rim. The latter was then increased by the return of wet conditions during the early part of this interval after logging dated between 2500 and 1300 cal yrs BP.

At 3300 cal yrs BP, lower montane forest taxa values are partially offset by increases of *Syzygium, Maesa* and *Gnidia*-type, indicating the opening of the forest. This opening peaks at 2600 cal yrs BP with the increase of Urticaceae and Poaceae undiff. . The evolution of the vegetation was organized in two stages. Between 2500 and 1300 cal yrs BP, a small forest recovery takes place. Forest vegetation is dominated by *Ilex mitis* associated with *Schefflera* and to a lesser extent *Sygyzium, Rapanea, Maesa, Nuxia,* and *Gnidia*-type. *Podocarpus* and *Olea capensis* are poorly represented. In the later part of this phase, the presence of sub-montane elements successively occurs such as *Macaranga, Celtis,* and *Lophira.* Finally, the Afromontane elements (Ericaceae undiff., *Hyperycum, Isoglossa*), with smaller percentages are generally better represented than in the high forest phase. This phase corresponds to wet period as suggested at Bambili by Stager and Anfang-Sutter (1999) who noted a P / E positive. The lake level rose during this period as reflected in the diatom assemblages; however, this could be artificially enhanced by the increased sediment transport from the lake margin. A humid climate is also noted in Ossa between 2700 and 1300 cal yrs BP by Nguetsop et al. (2004) confirming its regional character. At 960 cal yrs BP, the deep decline of *Schefflera,* partially offset by higher percentages of *Olea,* illustrates another phase of environmental degradation. The opening in the middle is highlighted by the dominance of Poaceae and Urticaceae. Between 960 cal yrs BP to the present, a marked upturn in forest is observed, with expansion of *Schefflera* (45%). The degradation of forest leads to a generalized fall in the majority of taxa, montane and submontane. The presence of Asteraceae undiff, *Ilex*

Vegetation Evolution in the Mountains of Cameroon During the Last 20,000 Years:
Pollen Analysis of Lake Bambili Sediments

131

mitis, Ficus, Embelia-type, *Maesa, Rapanea, Sygyzium, Gnidia*-type, *Clematis*-type, *Nuxia*-type, *Hypoestes*-type, *Alchornea*, Sapotaceae undiff., Solanaceae undiff, however, *Celtis* reveals nature montane forest mixed with the semi-deciduous forest.

5.6 The impact of environmental change on biodiversity

Fluctuations in biodiversity, calculated using the method developed by Birks and Line (1992) (Figure 8), show four phases of maximum biodiversity that correspond to periods of transition or disturbance of the forest. These phases are centered around 18,000 cal yrs BP, 12,000 cal yrs BP, 6500 cal yrs BP and 1500 cal yrs BP. The highest values of the record occur during the post-glacial colonization of forest following the YD (ca 12,000 cal yrs BP) and at the end of the small logging period (ca 1000 cal yrs BP). During these periods, forest biodiversity is increased due to the internal forest dynamics incorporating the gradual disappearance and appearance of certain plants resulting in greater overall species richness. According to Birks and Line (1992), floristic richness is favored during phases of "intermediate" disruption. The ecosystem fragmentation associated with such disruption limits the domination of one single component as long as the disturbance is not sufficient to cause the complete disappearance of the ecosystem on a regional scale.

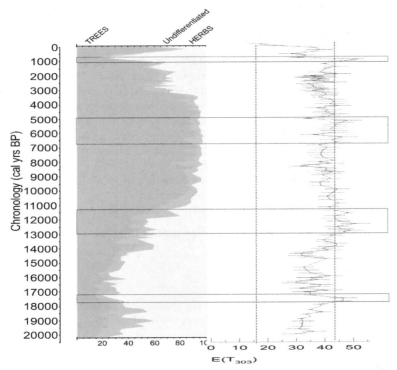

Fig. 8. Physiognomy of the vegetation and changes in biodiversity at Bambili. Left, Bambili diagram synthesis with the percentages of trees, herbs and undifferentiated. Right, rarefaction analysis showing Bambili biodiversity. Biodiversity is very high in the transition phases of the vegetation (red boxes). The dashed lines indicate the minimum of biodiversity (left line) and the value of average biodiversity (right line).

At Bambili, highest biodiversity does not follow forest stabilization, in contrast, biodiversity indices are generally low during these times. As shown in Figure 8, the current biodiversity surrounding Lake Bambili is the lowest in the last 20,000 years. This confirms the considerable impoverishment of the environment related to the recent deterioration of the forest environment undoubtedly amplified by the action of man, as evidenced by the comparative analysis of satellite photos from 1998 and 2003.

6. Conclusion

This palynological study of the paleoenvironments surrounding Bambili reveals the history of montane Central Atlantic Africa over the last 20,000 years. It also provides unique evidence on the response of Cameroon's mountain forests to climate change. During the LGM, a highly degraded forest formation was present around Bambili dominated by light-demanding components suggesting a dry environment. For the Holocene, the proportion of tree pollen shows continuous forest cover between 10,000 and 3300 yrs cal BP. The expansion of forest, very dynamic at the beginning of the Holocene, led to installation of a montane forest dominated by *Schefflera, Podocarpus* and *Olea* which responded individually to climate change. This forest lasted for much of the Holocene, then floristic composition changed. From this period, forest degrades very sharply during three centuries and the forest loses about 40% of its importance.

This is in agreement with the general context of equatorial forest evolution as a result of drier conditions of the end of the Holocene humid period. As emphasized by Birks Line (1992) "floristic richness is maximized by the disruption and fragmentation of the landscape when it reaches a level sufficient to prevent the domination of a single species and insufficient to cause the extinction of all components of the landscape. "The lowest percentages of trees on the top of the Bambili sequence dominated by a single taxon: *Schefflera*, indicates the disturbance of the landscape which reached a maximum level showing considerable impoverishment of the local flora. However, despite the effects of climate, intensified anthropogenic impacts have dramatically reduced forest biodiversity in recent decades. Continuation of such practices, associated largely with agriculture and ranching, in the future will likely lead to the disappearance of this ecosystem.

Palynological studies on other sites of Cameroon during LGM are needed to better assess the extent of *Podocarpus* forest.

7. Acknowledgements

This work was financed by IFORA and C3A. We thank for their contributions Anne-Marie Lézine, Emile Roche, Annie Vincens, Sarah Ivory, the LSCE and the palynology laboratory of the University of Liège.

8. References

Assi-Kaudjhis, C., 2011. Dynamique des écosystèmes et biodiversité des montagnes du Cameroun au cours des derniers 20 000 ans. Analyse palynologique d'une série

Vegetation Evolution in the Mountains of Cameroon During the Last 20,000 Years:
Pollen Analysis of Lake Bambili Sediments

133

sédimentaire du lac Bambili. Thèse de Doctorat. Université de Versailles, Université de Liège, 173p.

Assi-kaudjhis, C., Lézine, A-M. & Roche, E., 2008. Dynamique de la végétation d'altitude en Afrique centrale atlantique depuis 17000 ans BP. Analyses préliminaires de la carotte de Bambili (Nord-Ouest du Cameroun). Geo-Eco-Trop, 32: 131-143.

Birks, H.J.B. & Line, J.M., 1992. The use of rarefaction analysis for estimating palynological richness from Quaternary pollen-nalytical data.The Holocene 2, 1: 1-10.

Bonnefille, R., Riollet, G., Buchet, G., Icole, M., Lafont, R., Arnold, M., & Jolly, D., 1995. Glacial/Interglacial record from intertropical africa, High resolution pollen and carbon data et Rusaka, Burundi. Quaternary ScienceRreviews, Vol. 14, pp. 917-936.

Coetzee, J.A., 1967. Pollen analytical studies in East and Southern Africa. Palaeoecology of Africa 3: 1-146.

Demenocal, P., Ortiz, J., Guilderson, T., Adkins, J., Sarnthein, M., Baker, L. & Yarusinsky, M., 2000. Abrupt onset and termination of the African Humid Period: rapid climate responses to gradual insulation forcing. Quaternary Science Review 19: 347-361.

Dupont, L., M., Jahns, S., Marret, F. & Ning, S., 2000. Vegetation in equatorial West Africa: time-slices for the last 150ka. Palaeogeography, Palaeoclimatology,

Elenga, H. & Vincens, A., 1990. Paléoenvironnements quaternaires récents des plateaux Bateke (Congo) : étude palynologique des dépôts de la dépression du bois de Bilanko. Faunes, Flores, Paléoenvironnements continentaux. In : Paysages Quaternaires de l'Afrique Centrale Atlantique. ORSTOM : 271-282.

Elenga, H., Schwartz, D. & Vincens, A., 1994. Pollen evidence of late Quaternary vegetation and inferred climate changes in Congo. Palaeogeography, Palaeoclimatology, Palaeoecology 109: 345-356.

Faegri K. & Iversen, J., 1992. Textbook of Pollen Analysis. IV Edition. Printed and bound in Great Britain by Courier International Limited, East Kilbride; 328p.

FAO, 2002. Web LocClim, Local Monthly Climate Estimator (LocClim 1.0).

Farrera, I., Harrison., S.P., Prentice, I.C., Ramstein, G., Guiot, J., Bartlein, P. J., Bonnefille, R., Bush, M., Cramer, W., Grafenstein, Von, U., Holmgren, K., Hooghiemstra, H., Hope, G., Jolly, D., Lauritzen, S.-E., Ono, Y., Pinnot, S., Stute, M. & Yu, G. 1999. Tropical climates at the last Glacial Maximum: a new synthesis of terrestrial palaeoclimate data. Vegetation, lake-levels and geochemistry. Climate dynamics 15: 823-856.

Gasse, F. 2000. Hydrological changes in the African tropics since the Last Glacial Maximum. Quaternary Science Reviews 19, 1-5 : 189-211.

Grimm, E., 1987. CONISS: a fortran 77 program for stratigraphically constrained cluster analysis by the method of incremental sum of squares. Computers and Geosciences 13, 1: 13-35.

Hamilton, A.C., 1982. Environmental history of East Africa. A study of the Quaternary. Academic Press, London : 328 pp.

Jowsey, P.C., 1966. An improved peat sampler. New Phytol 65, 245-248.

Kadomura, H. & Kiyonaga, J., 1994. Origin of grassfields landscape in the West Cameroon Highlands. In: Kadomura H. (ed.). Savannization Processes in Tropical Africa II. Tokyo Metropolitan University, Japan: 47-85.

Kageyama, M., Combourieu Nebout, N., Sepulchre, P., Peyron, O., Krinner, G., Ramstein, G. & Cazet, J-P, 2005. Last Glacial Maximum and Heinrich Event 1 in terms of climate and vegetation around the Alboran Sea: a preliminary model-data comparison. C.R. Geoscience 337, 983–992.

Lebamba, J., 2009. *Relation pollen-végétatin-climat actuel en Afrique centrale. Une approche numérique appliqué à la sequence quaternaire du lac Barombi Mbo, Cameroun*. Thèse Unviersité Montpellier II : 253p.

Letouzey, R., 1968. *Etude phytogéographique du Cameroun*. Encyclopédie Biologique, 69, 511 p.

Letouzey, R., 1985. *Notice de la carte phytogéographique du Cameroun au 1/500.000è*. Inst. Carte Intern. Végétation, Toulouse, et Inst. Rech. Agron., Yaoundé, 240 p.

Lézine, A-M., 1997. Evolution of the west African mangrove during the late Quaternary: a review. *Geographie physique et Quaternaire* 51, 3: 405-414.

Lézine, A-M. & LE Thomas, A., 1995. Histoire du massif forestier ivoirien au cours de la dernière déglaciation. *In:* 2è Symposium de Palynologie africaine, Tervuren (Belgique) / 2nd Symposium on African Palynology, Tervuren (Belgium). *Publ. Occas. CIFEG*, Orléans, 1995/31: 73-85.

Lézine, A-M. & Cazet, J-P., 2005. High-resolution pollen record from core KW31, Gulf of Guinea, documents the history of the lowland forests of West Equatorial Africa since 40 000 yr ago. *Quaternary Research* 64: 432-443

Lézine, 2007. Postglacial Pollen records of Africa. *In* Scott A Elias (ed.) *Encyclopedia of Quaternary Sciences*, Elsevier, 4, 2682-2699.

Livingstone, D. A., 1967. Postglacial vegetation of the Rwenzori Mountains in Equatorial Africa. *Ecological Monographs* 37, 1: 25-52.

Maley, J., 1987. Fragmentation de la Forêt dense humide africaine et extension des biotopes montagnards au Quaternaire récent: nouvelles données polliniques et chronologiques. Implications, paléoclimatiques, chronologiques et biogéographiques. *Paleoecology of Africa* 18: 307-334.

Maley, J. & Livingstone , A., D., 1983. Extension d'un élément montagnard dans le sud du Ghana (Afrique de l'Ouest) au Pléistocène supérieur et à l'Holocène inférieur: premières données polliniques. *C. R. Acad. Sc. Paris*, 296, Série II : 1287-1292.

Maley, J., 1992. Mise en évidence d'une péjoration climatique entre ca 2 500 et 2 000 ans B.P. en Afrique tropicale humide. *Bull. Soc. géol. France* 163, 3 : 363-365.

Maley, J. & Brenac, P., 1998. Vegetation dynamics, palaeoenvironments and climatic changes in the forests of western Cameroon during the last 28 000 years BP. *Review of Palaeobotany and Palynology* 99 : 157-187.

Mix, A., C., Bard E & Schneider, R., 2001. Environmental processes of the Ice Age: land, oceans, glaciers (EPILOG). *Quaternary Science Review* 20 : 627-657.

du Mont Oku (Cameroun). Thèse en Biologie-Santé, Univ. de Picardie et Univ. de Youndé 1 : 161p.

Marzin, C. & Braconnot, P., 2009. The role of the ocean feedback on Asian and African monsoon variations at 6 kyr and 9.5 kyr BP. *C. R. Geoscience* 341 : 643–655.

Momo Solefack, M. C., 2009. *Influence des activités anthropiques sur la végétation du Mont Oku (Cameroun)*. Thèse en Biologie-Santé, Univ. de Picardie et Univ. de Youndé 1 : 161p.

Mumbi, C., T., Marchant, R., Hooghiemstra , H. & Wooller, M.J., 2008. Late Quaternary vegetation reconstruction from the Eastern Arc Mountains, Tanzania. *Quaternary Research* 69: 326-341.

Nguetsop, V.F., Servant-Vildary, S., Servant, M. & Roux, M, 2004. Long and short-time scale climatic variability in the last 5500 years in Africa according to modern and fossil diatoms from Lake Ossa (Western Cameroon). *Glob. Planet. Change* 72: 356-367.

Pissart, A. 2002. Concernant la disparition du Gulf Stream pendant la dernière glaciation et le danger de voir se reproduire ce phénomène catastrophique pour l'Europe. *Bulletin de la Société Géographique de Liège*, 42 : 79-83.

Puig, H., 2001. *La Forêt tropicale humide*. Belin, 448 p.

Reynaud-Farrera, I., 1995. *Histoire des paleoenvironnements forestiers du Sud Cameroun à partir d'analyses palynologiques et statistiques de dépôts holocènes et actuels*. Thèse, Univ. De Motpellier II Sciences et Techniques du Languedoc : 198p.

Richards, K., 1986. Preliminary results of pollen analysis of a 6000 year core from Mboandong, a crater lake in Cameroun. *In*: Baker R.G.E., Richards K., Rimes C.A. (Hrsg.): The Hull University. Cameroun Expedition 1981-1982. Final Report. Department of Geography, University of Hull: 14 -28.

Roberts, N., Taieb, M., Barker, P., Damnati, B., Icole M.& Williamson, D., 1993. Timing of the Younger Dryas event in East Africa from lake-level changes. *Nature* 366 : 146-148.

Roberts, N., L., Piotrowski, A., M., Mcmanus, J., F.& Keigwin, L., D.,, 2010. Synchronous Deglacial Overturning and Water Mass Source Changes. *Science* 327 : 75-77.

Salzmann, U., 2000. Are modern savannas degraded forests? - A Holocene pollen record from the Sudanian vegetation zone of NE Nigeria. *Vegetation History and Archaebotany* 9: 1-15.

Sarnthein, M., Tetzlaff, G., Koopmann, B., Wolter, K. & Pflaumann, U., 1981. Glacial and interglacial wind regimes over the eastern subtropical Atlantic and North-West Africa. *Nature* 293: 193–296.

Shanahan, T., M., Overpeck, J,. T., Wheeler, C., W., Beck, J., Pigati, W., J. S., Talbot, M. R., Sholz, C. A., PECK, J. & KING, J. W., 2006. Paleoclimatic variations in West Africa from a record of late Pleistocene and Holocene lake level stands of Lake Bosumtwi, Ghana. *Palaeogeography, Palaeoclimatology, Palaeoecology* 242: 287-312.

Schröter, D., Cramer, W., Leemans, R., Prentice, I., C., Araújo, M., B., Arnell, N., W., Bondeau, A., Bugmann, H., Carter, T., R., Gracia, C., A., De La Vega-Leinert, A., C., Erhard, M., EWERT, F., Glendining, M., House, J., I., Kankaanpää, S., Klein, R., J., T., Lavorel, S., Lindner, M., Metzger, M., J., Meyer, J., Mitchell, T., D., Reginster, I., Rounsevell, M., Sabate, S., Sitch, S., Smith, B., Smith, J., Smith, P., Sykes, M., T., Thonicke, K., Thuiller, W., Tuck, G., Zaehle, S. & Zierl B., 2005. Ecosystem Service Supply and Vulnerability to Global Change in Europe. *Science* 310: 1333-1337.

Stager C.J. & Anfang-Sutter, R. 1999. Preliminary evidence of environmental changes at Lake Bambili (Cameroon, West Africa) since 24 000 BP. *Journal of Paleolimnology* 22: 319-330.

Stuiver, M., REIMER, P.J., & REIMER, R., W., 2005. CALIB 5.0. [program and documentation]. http://calib.qub.ac.uk/calib/

Suchel, J-B, 1988. *Les climats du Cameroun*. Thèse de doctorat d'Etat, Université de Saint-Etienne : 789 p.

Tamura, T., 1990. Late Quaternary landscape evolution in the West Cameroon Highlands and the Adamaoua Plateau. *In*: Lanfranchi, R. & Schwartz, D. (Eds). Paysages quaternaires de l'Afrique centrale atlantique. *ORSTOM*., Paris: 298-318.

Timmermann, A., & Menviel, L., 2009. What Drives Climate Flip-Flops? *Science* 325, 273-274.

Thomas, C., D., Cameron, A., Green, R., E., Bakkenes, M., Beaumont, L., J., Collingham, Y., C., Erasmus, B., F., N., Ferreira De Siqueira, M., Grainger, A., Hannah, L., Hughes, L., Huntley, B., Van Jaarsveld, A., S., Midgley, G., F., Miles, L., Ortega-Huerta, M., A., Townsend Peterson, A., Phillips, O., L., & Williams, S., E., 2004. Extinction risk from climate change. *Nature* 247:145-148.

Thuiller, W., Midgley, G., F., Hughes, G.O., Bomhard, B., Drew, G., Rutherford, M.C. & WO, O., 2006. Endemic species and ecosystem sensitivity to climate change in Namibia. *Global Change Biology* 12: 759–776.

Vincens, A., Lézine, A.-M., Buchet, G., Lewden, D., Le Thomas, A., and contributors, 2007. African Pollen Database inventory of tree and shrub pollen types, Rev. Palaeobot. Palyno., 145, 135–141.

Vincens, A., Garcin, Y. & Buchet, G., 2007. Influence of rainfall seasonality on African lowland vegetation during the late Quaternary : Pollen evidence from lake Masoko, Tanzania. *Journal of Biogeography* 34: 1274-1288.

Vincens, A., Buchet, G., Servant, M., & ECOFIT Mbalang collaborators, 2010. Vegetation response to the African Humid Period termination in central Cameroon (7°N) – new pollen insight from Lake Mbalang. *Clim. Past.Discuss.* 5: 2577-2606.

Von Grafenstein, U., Erlenkeuser, H., Müller, J., Jouzel, J. & Johnsen, S., 1997The cold event 8200 years ago documented in oxygen isotope records of precipitation in Europe and Greenland. *Clim. Dyn.* 14: 73-81.

Walther, G-R., Post, E., Convey, P., Menzel, A., Parmesank, C., Beebee, T., J,. C., Fromentin, J-M., Hoegh-Guldbergi, O., & Bairlein, F., 2002. Ecological responses to recent climate change. *Nature* 416: 389-395.

Watrin, J., Lézine, A.M., Hély, C. & Contributors, 2009. Plant migration and plant communities at the time of the "green Sahara"; External geophysics, climate and environment; *C. R. Geoscience* 341: 656–670

White, F., 1983. *The vegetation of African: a descriptive memoir to accompany the UNESCO/AETFAT/UNSO vegetation map of Africa.* Paris, France: UNESCO.

White, F., 1986. *La végétation de l'Afrique. Recherche sur les ressources naturelles* XX. ORSTOM – UNESCO: 384p.

Zogning, A., Giresse, P., Maley, J. & Gadel, F., 1997. The Late Holocene palaeoenvironment in the Lake Njupi area, West Cameroon: implications regarding the history of Lake Nyos. *Journal of African Earth Sciences* 24: 285-300.

6

Using Remotely Sensed Imagery for Forest Resource Assessment and Inventory

Rodolfo Martinez Morales

International Crops Research Institute for the Semi-Arid Tropics, Niamey
Niger

1. Introduction

Forests are complex ecosystems that develop over centuries through the interactions between organisms and biogeochemical cycles of elements occurring in the soil-atmosphere continuum. The biomass and structure of a forest stand is involved in several ecosystem processes and has been used as an indicator of forest health and productivity. The forest biomass is a key component of the carbon cycle, as forests represent large carbon sources and sinks (Skole & Tucker, 1993). Tree canopy height and area are highly correlated with biomass and are important inputs in forest productivity models (Drake, 2001). The variation of forest biomass production has been related to variations in canopy light absorption since the amount and spatial distribution of vegetation, directly affects light availability in forests. Forest stand factors that determine light absorption include: amount of leaf area, crown and canopy structure, phenology, and leaf optical properties (Jarvis and Leverenz, 1983). The amount of leaf area, measured through the leaf area index (LAI), is considered a key parameter of ecosystem processes (Asner and Wessman, 1997). Several forest ecosystem processes are strongly controlled by LAI including interception of light (Machado & Reich, 1999) and precipitation (Van Dijk & Bruijnzeel, 2001), gross primary productivity (Jarvis & Leverenz, 1983), transpiration (Granier et al., 2000), and soil respiration (Davidson et al., 2002). LAI is also related to other important ecological processes such as evapotranspiration, CO_2 and water exchange with the atmosphere, nutrient cycling and nutrient storage in plants (Dougherty et al. 1995). Therefore, measurements of forest biomass and structure are critical in the study of ecosystems for many applications including management of forest plantations, wildlife and biodiversity, fire modeling, and carbon sequestration among others.

Traditionally, the assessment of forest structure and growth has been done by measuring forest canopy attributes such as tree canopy dimensions, height and LAI in the field using hand-held equipment including leaf area meters, height poles, clinometers and measure tapes. Although field-based methods can be highly accurate, they are typically limited in scope to either mapping at plot scales or sampling sites at the landscape scale. Because of the expense of conducting detailed forest inventories over large areas, considerable research has focused on developing tools to estimate forest canopy attributes using remote sensing techniques. Historical aerial photos have proven useful, but analysis has generally been done manually. With new satellite sensors and improved computing power and analytical software, remote sensing is becoming an important tool for forest cover mapping, environmental monitoring, and ecological process assessments from global, regional, and landscape levels (Plummer, 2000).

This chapter is a review of the remote sensing technologies currently used to achieve more accurate forest resource inventory and assessment at landscape and regional scales. The review had three main objectives. The first objective was to describe remote sensing principles and technology development for forest research worldwide. The second objective was to present practical applications of remote sensing technologies used to characterize forest structure and health at the stand and individual-tree levels. The third objective was to present strategies for effective use of remote sensing technologies to improve management of forests worldwide.

2. Remote sensing technologies for forest research

The basic principle in using remote sensing is based on the selective nature of radiation absorption by vegetation canopies, resulting in unique spectral signatures that describe distinctive patterns of short-wave (visible and infrared) radiation reflectance. The reflectance spectrum of green vegetation is characterized by low reflectance in the red region (0.6 - 0.7 μm), associated with chlorophyll absorption, and strong near infrared (NIR) reflectance (0.7 -1.2 μm) related to internal leaf structure (Jensen, 2000; Roberts et al., 1997). Satellites may be either active or passive and are designed to capture reflectance from various regions of the electromagnetic spectrum as multispectral bands. While active satellite sensors transmit signals which are detected or emitted back at the instrument after hitting the earth surface, passive sensors do not transmit energy signals, but rather only detect reflected energy from earth in the visible and infrared regions. Available multispectral satellite imagery from passive sensors over the last 30 years and improved satellite imaging technologies over the last 10 years have increased the capabilities to describe spatial and temporal dynamics of vegetation characteristics at numerous scales. Multispectral imagery from medium resolution sensors, such as Landsat (30-m pixel resolution) (Curran et al., 1992, Baugh and Groeneveld, 2006; Xu, 2007) and Spot (5-m pixel resolution) (Soudani et al., 2006), have been used to assess vegetation conditions and phenological changes in forested areas at regional and landscape scales. Günlü et al. (2008) integrated the analysis of Landsat imagery with conventional forest inventory measurements and ecological and physiographic information to produce site quality index maps for various temperate forest species in Turkey. However, the spatial resolution of medium resolution satellites does not allow resolving forest stands and individual trees. Fine resolution satellites such as Ikonos, QuickBird, and GeoEye1 have increased pixel resolution down to less than one meter for panchromatic images and 2-4 meters for multispectral images capturing the blue, green, red and NIR spectral regions (Table 1). The analysis of this imagery has provided a way to study large areas by allowing visualization of entire landscapes and regions and identification of individual tree species. Due to high temporal frequency of flights over the same area (3 to 4 days), fine resolution satellites have facilitated assessments of forest structure, condition and health across multiple spatial and temporal scales. Imagery from these satellites has improved the identification and mapping of individual forest species across entire landscapes. While the high spatial resolution allows for delineation of single tree crowns, the multispectral bands allow for determination of variations of canopy greenness within forest stands (Guo et al., 2007). In particular, these satellites have been successfully applied for forest inventory in tropical environments and allowed for the mapping of tree crown sizes (Martinez Morales et al., 2008), tree density, species identification, and assessment of temporal changes in individual tree growth and mortality (Clark et al., 2004; Martinez Morales et al., 2011).

Satellite	Multispectral	Pancromatic
Ikonos	4	1
QuickBird	2.62	0.65
WorldView2	2	0.65
GeoEye1	2	0.5

Table 1. Pixel resolution in meters for common fine resolution satellites.

While high spatial resolution satellite sensors can be used to assess forest structural characteristics, they only collect data on a limited number of spectral bands (blue, green, red, and near-infrared). Hyperspectral remote sensing, or imaging spectroscopy, collects data on hundreds of bands from visible to infrared wavelengths (0.4 to 2.5 μm). Due to higher definition of unique spectral signatures among vegetation types, this kind of data has expanded the potential to identify forest species and assess canopy biochemical and physiological properties such as leaf pigments, carbon and nitrogen content at the species level (Asner et al., 2005; Clark et al., 2005; Pu et al., 2008; Féret and Asner, 2011). Imagery from Hyperion EO-1, a hyperspectral satellite which detects 220 spectral regions at 30-m pixel resolution (eo1.usgs.gov), has been used to map structural vegetation metrics and indices of forest productivity at regional scales (Pu and Gong, 2004; Asner et al., 2005). However, the spatial resolution of this satellite does not allow resolving forest stands and individual trees. The Airborne Visible and Infrared Imaging Spectrometer (AVIRIS) built by NASA (aviris.jpl.nasa.gov) provides 224 contiguous spectral bands with pixel resolution varying from 2 to 20-m depending on flight altitude. These data have improved remotely sensed predictions of forest health, biomass, species identity and variation through a better understanding of spectral responses of forest canopies at the species level (Roberts et al., 1997; Asner and Lobell, 2000). Asner et al., 2008 used AVIRIS data to analyze the reflectance properties of 37 distinct species (7 common native and 24 introduced tree species) in order to spectrally differentiate between native and alien trees in a montane forest in Hawaii. They found that the reflectance signatures of Hawaiian native trees were unique from those of introduced trees. Since the AVIRIS imaging system is costly and frequently unavailable, a number of companies such as Hughes, Lockheed and Surface Optics among others, have developed a variety of visible and infrared imaging spectrometers available in the market. Greg Asner's research team has pioneered the use of airborne hyperspectral sensors to extract detailed biochemical data on plant canopies in Hawaii's forests. Distinct structural or biochemical signatures have been used to map the distribution of native forest species and several tree and shrub invasive species (Asner et al., 2008a; Asner et al., 2008b).

Although passive satellite sensors offer routine and repeated assessments at scales down to 1 meter, this technology has difficulty in capturing reflectance beyond upper canopy layers and is better suited for mapping horizontal structure rather than vertical structure (Weishampel et al., 2000). Active remote sensing technologies offer great potential to spatially map a forest three-dimensional (3D) structure at various scales from landscape, stand and individual tree levels. Active satellite systems based on interferometric synthetic aperture radar (InSAR) can provide measures of horizontal and vertical structure of vegetation at regional scales (Treuhaft & Siqueira, 2000), but this technology does not provide the spatial resolution required in detailed forest studies. However, active airborne laser scanning sensors such as LIDAR (Light Detection and Ranging System) are providing improved capabilities for the estimation of forest canopy dimensions at the individual tree level (Weishampel et al., 2000; Hyde et al., 2005; Chen, 2006). Small foot-print LIDAR systems have provided 3D surveys of forest canopy and have resolved some of the challenges not met by existing techniques for measuring canopy structure (Hetzel et al., 2001; Tickle *et al.*, 2006; Chen et al., 2006). The isolation and extraction of tree structural information from LIDAR imagery has allowed more explicit ecological modeling through the estimation of individual-tree height, crown area, trunk height, biomass and leaf area (Henning, 2005; Chen et al., 2007, Chen, 2010). Chen et al., 2006 isolated trees from small-footprint airborne LIDAR data in deciduous oak woodland in California using a marker-controlled watershed segmentation method and a canopy height model derived from the LIDAR data. In the same site, Chen et al., 2007 proposed a new metric called canopy geometric volume derived from LIDAR data to estimate individual tree height, crown size, LAI, basal area and stem volume at 70 % accuracy. On Hawaii montane forests, Asner et al., 2009 derived canopy vertical profiles from LIDAR imagery in order to quantify 3-D forest structure and above ground biomass (AGB). They found that LIDAR measurements were strong predictors of AGB ($R^2 = 0.78$) across sites and species. Combining or fusing the highly detailed vertical measurements provided by LIDAR and the broad-scale mapping capabilities of passive optical sensors can provide dramatic increases in forest mapping and characterization. Wulder et al., 2004 used texture metrics from Landsat images to improve LIDAR estimates of canopy height. Hyde et al. 2006, combined forest structural information from LIDAR and QuickBird to improve estimates of canopy height and biomass. Asner et al. (2008a) combined airborne LIDAR and hyperspectral imagery to differentiate and map native and alien tree species in Hawaii montane forests, including understory plants like Kahili ginger (*Hedychium gardnerianum*) and strawberry guava. Therefore, airborne systems combining LIDAR with hyperspectral sensors have the highest potential for reliable estimations of individual-tree structure parameters such as canopy size, volume and leaf area.

However, airborne LIDAR imaging systems have disadvantages due to the high cost of flight time and a large number of flights for imaging entire landscapes. Resource Mapping Hawaii Inc (www.remaphawaii.com), developed a system for mapping detailed forest structural and morphological characteristics using ultra high resolution airborne multispectral imagery at 1.5 cm per pixel. The Nature Conservancy of Hawaii is employing such imaging system to map the distribution of Australian tree fern. Geo-referenced locations of individual trees can be obtained from this imagery and uploaded to a handheld GPS, allowing for more efficient eradication efforts (Ambagis et al., 2009). This imaging technology best complements to field inventories, providing detailed information of vegetation in areas that are remote, inaccessible, or rapidly changing.

3. Remote sensing applications in forest research

With the development of advanced image processing techniques, remote sensing technology has rapidly expanded to allow estimation of forest cover in heterogeneous landscapes and estimation of tree density, species identification and assessment of temporal changes in individual tree growth, health and mortality across entire landscapes (Carleer and Wolff, 2004; Carleer and Wolff, 2005; Clark et al., 2004; Chubey et al., 2006; Soudani et al., 2006). Martinez Morales et al., 2011 developed practical methodologies to analyze fine resolution satellite imagery using pixel-based image classification techniques for forest resource assessment. They fused GeoEye1 multispectral and panchromatic bands to conduct landscape-level assessments of koa (*Acacia koa*) forest health across an elevation range of 600–1,000 m asl in the island of Kauai. The goal of the study was to assess the spatial distribution of koa forest dieback patterns across a gradient of temperature and rainfall in order to determine the influence of these environmental factors on dieback patterns. The spectral bands were analyzed using a supervised classification technique to differentiate and classify pixels representing healthy and unhealthy koa stands and other land cover classes existing in the landscape. They classified healthy koa forest stands at 87 % accuracy from areas dominated by introduced tree species and differentiated healthy koa stands from those exhibiting dieback symptoms at 98 % accuracy. A landscape-scale map of healthy koa forest and dieback distribution (Fig. 1) demonstrated larger presence of unhealthy koa stands in areas with lower elevation and precipitation and higher temperature.

While pixel-based image classification involves assigning individual pixels to a vegetation class according to unique reflectance patterns across the spectral bands (spectral signature), object-based methods also include class shape and texture as additional parameters (Jensen 2000). Object-based analysis and image segmentation techniques have been increasingly applied in fine resolution multispectral imagery as an alternative to overcome the difficulties of conventional procedures of spectral image analysis for various forestry applications (Chubey et al., 2006; Herold et al., 2003; Hu et al., 2005). Instead of analyzing a single pixel spectral response, a wide range of spectral values in a group of pixels representing a forest stand is interpreted as a homogeneous object which can be further segmented into even more homogeneous subgroups. Pixel grouping can be controlled by the user through the definition of parameters such as size, homogeneity and shape in order to reduce heterogeneity in the resulting objects (Chubey et al., 2006). Wang et al. (2004) utilized a combination of spectral classification techniques and segmentation methods for tree-top detection and tree classification in a forested area in British Columbia, Canada. They calculated the first principal component from a set of spectral images from the Compact Airborne Spectrographic Imager and applied a Laplacian edge detection method for tree-crown delimitation. They further applied a segmentation technique and tree-top markers in order to differentiate final individual tree crowns at 85% accuracy. In a Belgian forest, Kayitakire et al. (2006) found highly significant relationships between image texture metrics extracted from the IKONOS panchromatic band with several forest productivity indices including tree density, height, crown size and basal area. Since the IKONOS NIR band contains important vegetation information, Herold et al. (2003) used this band to derive various texture and landscape metrics that classified forests at 78 % accuracy along a California coast region. Martinez Morales et al. (2008) analyzed fused Ikonos multispectral and panchromatic bands with spectral and object-based classification methods, to estimate forest cover at 86% accuracy in a Hawaiian dry forest ecosystem. Their comparison between spectral and object-based methods demonstrated superior performance of object-based classification algorithms in delineating tree canopy cover in a highly heterogeneous dry forest environment. The object-based approach allowed for differentiation of tree crowns, tree shades, and their transitional areas from other objects of similar size, shape, or spectral range such as green grass and lava outcrops (Fig. 2). A particular important result was the clear delimitation of individual tree crown areas that can be useful for forest inventory even on high spatial heterogeneity of vegetation conditions.

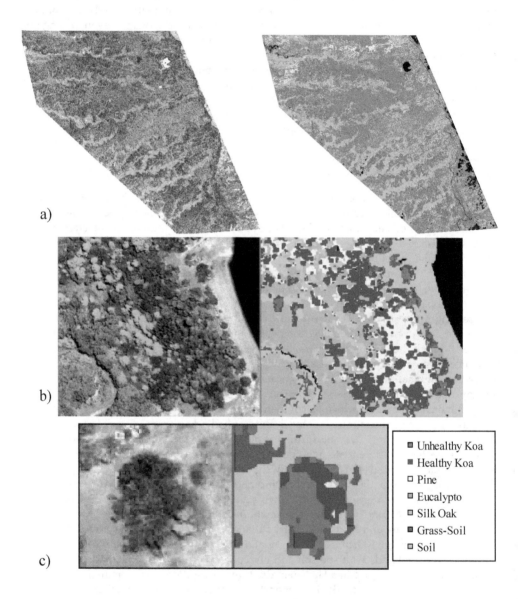

Fig. 1. A montane forest ecosystem from the Island of Kauai as viewed by the GeoEye1 satellite: a) Natural color composite at 0.5-m pixel resolution (left) and its corresponding classification (right); b) Image close-up depicting clear differentiation among tree species; c) Detailed close-up showing classification of diseased from healthy forest stands (Martinez Morales et al., 2011).

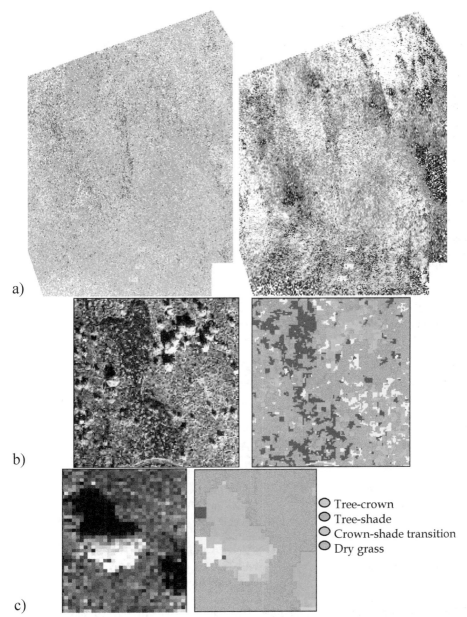

Fig. 2. A dry forest ecosystem from the north Kona region in the Island of Hawaii as viewed by the Ikonos-2 satellite. a) Natural color composite at 1-m pixel resolution (left) and its corresponding classification (right); b) Image close-up depicting differentiation among objects with similar reflectance (tree crowns from shrubs and grasses and tree shades from lava outcrops); c) Detailed close-up showing clear delineation of individual tree crowns and tree shades (Martinez Morales et al., 2008).

Since the reflectance of green vegetation is low in the red region due to chlorophyll absorption, and strong in the NIR due to internal leaf structure, a number of vegetation indices (VIs) (Table 2) have been calculated using these two regions of the reflectance spectrum for assessments of vegetation biomass, chlorophyll abundance and light absorption (Baugh and Groeneveld 2006), phenological changes in forested areas (Loveland et al., 2005) and for detailed identification of forest tree species (Soudani et al. 2006, Martinez Morales et al., 2012). Roberts et al. (1997) successfully used the Normalized Difference Vegetation Index (NDVI) from AVIRIS hyperspectral imagery to estimate LAI and canopy cover at moderate scales in a California forest. Carleer and Wolff (2004) derived NDVI, principal components (PCs) and texture metrics from Ikonos satellite data and used them in the identification of tree species in a forested area in Belgium. Seven tree species, including two different ages, were successfully identified with 86% overall classification accuracy. At various forest stands of tenths of hectares, Soudani et al. (2006) used five different VIs, such

Vegetation index	Formulae Source
1. Simple Ratio	$SR = NIR / R$ (Jordan 1969)
2. Normalized Difference Vegetation Index	$NDVI = NIR - R / NIR + R$ (Rouse et al. 1973)
3. Soil Adjusted Vegetation Index	$SAVI = (1 + L) * (NIR - R) / (NIR + R + L)$ $L = 0.5$ (canopy background adjustment factor) (Huete 1988)
4. Atmospherically Resistant Vegetation Index	$ARVI = (NIR - R) / (NIR + Q_{RB})$, $Q_{RB} = R - \gamma (B - R)$, $\gamma = 1$ (calibration factor) (Kaufman and Tanré 1992)
5. Modified Soil Adjusted Vegetation Index	$MSAVI = (1 + L) * (NIR - R) / (NIR + R + L)$ $L = 1 - 2a * NDVI * WDVI$ (Qi et al. 1994)
6. Enhanced Vegetation Index	$EVI = G * (NIR - R) / (NIR + C1*R - C2*B + L)$ with $G = 2.5$, $C1 = 6$, $C2 = 7.5$, $L = 1$ (Liu and Huete 1995)
7. Modified Simple Ratio	$MR = (NIR / R - 1) / ((NIR / R)^{1/2} + 1)$ (Chen 1996)

Table 2. Spectral vegetation indices. R, NIR and B are red, near-infrared, and blue bands, respectively. For Modified Soil Adjusted Vegetation Index, WDVI = NIR - aR (a = 0.08, slope of the soil line). For Enhanced Vegetation Index, G, C1, C2 and L are coefficients to correct for aerosol scattering, absorption, and background brightness.

as NDVI, Soil Adjusted Vegetation Index (SAVI), Atmospherically Resistant Vegetation Index (ARVI), Enhanced Vegetation Index (EVI) and Simple Ratio (SR) calculated using data from the IKONOS and SPOT satellites to accurately classify various forest stands in France. They found that ARVI, NDVI and SR had similar and better predictions of LAI compared to SAVI and EVI. Kayitakire et al. (2006), estimated forest productivity at regional and landscape scales by relating VIs with LAI. This relationship has been used as a strong diagnostic tool to make silvicultural management recommendations (Flores 2006). Flores (2006) developed empirical models that were not affected by site, stand structure or time of the year to estimate LAI in broad areas of southern loblolly pine stands in USA using NDVI and SR from Landsat data and airborne hyperspectral data. Asner et al., 2005 found that the canopy water content index (NDWI) calculated from EO-1 data was superior than NDVI in capturing climate driven variations in canopy structure of a Hawaiian forest. In a Hawaiian koa forest, Martinez Morales et al., 2012 used Ikonos multispectral imagery to calculate six VIs (ARVI, EVI, NDVI, SAVI, SR, Modified Soil Adjusted Vegetation Index (MSAVI) and Modified Simple Ratio (MSR)) as a measure of vegetation greenness, and related those to biophysical measures of forest productivity such as tree height, basal area, leaf area index and foliar nutrients for spatial prediction at the landscape scale. This procedure allowed a clear differentiation of koa stands from areas dominated by grasses, shrubs, and bare lava. Vegetation indices allowed differentiation of three koa forest stand classes at upper, intermediate and lower elevations. In agreement with the image classification, analysis of variance of tree height and leaf phosphorus suggested there were also three significantly different groups of koa stands at those elevations.

4. Conclusions

Fine spatial resolution remote sensing allows not only visual interpretations of forest species but also automated classification of forest stands. Since the electromagnetic radiation captured by satellites has interacted with forest canopies through chemical absorption or physical scattering, it contains information about the chemical and physical properties of each vegetation type in the landscape. Therefore, the analysis of spectral data allows distinguishing not only forests species but also forest structural variations based on their unique reflection properties across the electromagnetic spectrum. Based on canopy greenness, analysis of these imagery can also be used to differentiate diseased from healthy forest stands. Such applications should improve forest inventory and collection of forest attributes for productivity assessments among forest scientists, decision-makers, and the general public involved in the ecological restoration, conservation and silviculture of important tree species worldwide.

Although field measurements for forest resource inventory and assessment are more accurate than satellite measurements, satellites collect data across broad areas, sample the full range of variation in forest metrics, capture broad trends and dynamic change in forest stands and help expand our understanding of forests beyond the plot level. As such, satellite data allow for integration across ground measurements, extending them across landscapes and regions and allowing detection of spatial and temporal changes in forests that we could not measure using conventional survey methods. Therefore, the analysis of satellite imagery has become a practical necessity to measure and manage forests at landscape and regional scales. The greatest strengths of satellite imagery are their monthly to daily frequency and view of entire regions, which could improve monitoring and verification of forest management for sustainable harvest and carbon sequestration. Aerial photos have proven useful, but the technology is costly and limited to small areas. The advent of high spatial resolution satellites such as Ikonos, Quickbird and GeoEye1 has changed the cost and availability of high-resolution imagery. If available for an area, archived imagery from these satellites can be acquired for one third of the original cost. With current technology, fine resolution remote sensing is suited to differentiate among forest species with classification accuracy usually decreasing with an increasing number of forest classes. It is also difficult to distinguish forests of different ages or composition, and primary forests from tree plantations and older secondary forests. However, remote sensing is rapidly developing by technological advancements in data gathering and processing. The GeoEye2 satellite at 0.25 m pixel resolution will be launched in 2012 and it is expected to revolutionize the management of forest ecosystems worldwide since it will allow more accurate assessment of the small-scale forest variability across environmental gradients. Improved characterizations and delimitations of forest species, stands types and growth stages along environmental gradients will allow development of more efficient silvicultural management practices according to site-specific ecological requirements.

The integrated analysis of environmental data with remote sensing imagery in a Geographical Information System (GIS) framework, allows inferences on how environmental factors influence forest ecosystem functioning. The increasing integration of GIS and remote sensing has facilitated display and communication of satellite imagery between scientists and the general public, as witnessed by the explosive growth in mapping using tools like Google Earth. Overall, remote sensing technologies are proving to be powerful research and management tools for the inventory and assessment of forests around the world. We are now at the point where both satellite and airborne sensing systems can provide reliable and detailed information at the individual-tree level. These technologies will become increasingly important for assessment and management of forests worldwide as we continue to face the challenges of land use pressures, invasive species, and climate change.

5. Acknowledgements

As this chapter contains information derived during my PhD studies at the Department of Natural Resources and Environmental Management, University of Hawaii at Manoa, I greatly appreciate the enormous support of my advisors Dr. Travis Idol and Dr. James B. Friday for sharing their knowledge and experience on the study of forest ecosystems.

6. References

Ambagis, S., Menard, T. and Schlueter, J. (2009). Remote sensing and invasive weed management. *Proc. Hawaii Conservation Conference*, Honolulu, Hawaii.
Asner, G.P. and Lobell, D.B. (2000). A biogeophysical approach for automated SWIR unmixing of soils and vegetation. *Remote sensing of environment*, 74: 99 – 112.
Asner, G.P. and C.A. Wessman (1997). Scaling PAR absorption from the leaf to landscape level in spatially heterogeneous ecosystems. *Ecological Modelling*, 103 (1): 81-97.

Asner, G.P., Carlson, K.M., Martin, R.E. (2005). Substrate age and precipitation effects on Hawaiian forest canopies from space imaging spectroscopy. *Remote sensing of environment*, 98:457-467.

Asner, G.P., Jones, M.O., Martin, R.E., Knapp, D.E. and Hughes, R.F. (2008a). Remote sensing of native and invasive species in Hawai'ian forests. *Remote Sensing of Environment*, 112, 1912–1926.

Asner, G.P., Knapp, D.E. Kennedy-Bowdoin, T., Matthew O. Jones, Martin, R.E., Boardman, J. and Hughes, R.F. (2008b). Invasive species detection in Hawaiian rainforests using airborne imaging spectroscopy and LiDAR. *Remote Sensing of Environment*, 112, 1942–1955.

Asner, G.P., Hughes, R.F., Varga, T.A., Knapp, D.E. and Bowdoin, T.K. (2009). Environmental and Biotic Controls over Aboveground Biomass Throughout a Tropical Rain Forest. *Ecosystems*, 12: 261–278.

Baugh, W.M. and D.P. Groeneveld (2006). Broadband vegetation index performance evaluated for a low-cover environment. International Journal of Remote Sensing 27, (21):4715–4730.

Carleer, A. P. and Wolff, E. (2004). Explotation of very high resolution satellite data for tree species identification. *Photogrammetric Engineering and Remote Sensing*, 70:135-140.

Carleer, A. P., Debeir, O. and Wolff, E. (2005). Assessmet of very high spatial resolution satellite image segmentations. *Photogrammetric Engineering and Remote Sensing*, 71:1285-1294.

Chen, J.M. (1996). Evaluation of vegetation indices and a modified simple ratio for boreal applications. *Canadian Journal of Remote Sensing*, 22, 229-242.

Chen, Q. (2010). Retrieving canopy height of forests and woodlands over mountainous areas in the Pacific coast region using satellite laser altimetry, *Remote Sensing of Environment*, 114, 1610-1627.

Chen, Q., Gong, P., Baldocchi, D. and Tian, Y.Q. (2007). Estimating Basal Area and Stem Volume for Individual Trees from Lidar Data. *Photogrammetric Engineering & Remote Sensing*, 73 (12):1-11

Chen, Q., Baldocchi, D.D., Gong, P. and Kelly, M. (2006). Isolating individual trees in a savanna woodland using small footprint Lidar data. *Photogrammetric Engineering and Remote Sensing* 72:923-932.

Chubey, M. S., Franklin, S. E. and Wulder, M. A. (2006). Object-based analysis of Ikonos-2 imagery for extraction of forest inventory parameters. *Photogrammetric Engineering and Remote Sensing* 72(4):383-394.

Clark, D.B., Read, J.M., Clark, M., Murillo Cruz, A., Fallas Dotti, M. and Clark, D.A. (2004). Application of 1-m and 4-m resolution satellite data to studies of tree demography, stand structure and land use classification in tropical rain forest landscapes. *Ecological Applications*, 14:16-74.

Clark, M., Roberts, D. A. and Clark, D. B. (2005). Hyperspectral discrimination of tropical rain forest tree species at leaf to crown scales, *Remote Sensing of Environment*, 96(3-4), 375-398.

Curran, P.J., J.L. Dungan, and H.L. Gholz (1992). Seasonal LAI in slash pine estimated with Landsat TM. *Remote Sensing of Environment* 39: 3 – 13.

Davidson, E.A., Savage, K., Bolstad, P., Clark, D.A., Curtis, P.S. and Ellsworth, D.S. (2002). Belowground carbon allocation in forests estimated from litterfall and IRGA-based soil respiration measurements. *Agricultural and Forest Meteorology*, 113, 39–54.

Dougherty, P.M., T.C. Hennessey, S.J. Zarnoch, P.T. Stenberg, R.T. Holeman, and R.F. Wittwer (1995). Effects of stand development and weather on monthly leaf biomass dynamics of a loblolly pine (Pinus taeda L.) stand. *Forest Ecology and Management*, 72: 213 – 227.

Drake, J. (2001). Estimation of tropical forest biomass. Dissertation thesis, University of Maryland, College Park, MD.

Féret, A. and Asner, G.P. (2011). Spectroscopic classification of tropical forest species using radiative transfer modeling. *Remote Sensing of Environment*, 115(9):2415-2422.

Flores, F.G., Allen, H.L., Cheshire, H.M., Davis, J.M., Fuentes, M. and D. Kelting (2006). Using multispectral satellite imagery to estimate leaf area and response to silvicultural treatments in loblolly pine stands. *Can. J. For. Res.* 36(6):1587-1596.

Granier, A., Ceshia, E., Damesin, C., Dufrêne, E., Epron, D. and Gross, P. (2000). The carbon balance of a young Beech forest. *Functional Ecology*, 14, 312–325.

Günlü, A., Zeki, E.B., Kadıoğulları, I.A. and Altun, L. (2008). Forest site classification using Landsat 7 ETM data: A case study of Maçka-Ormanüstü forest, Turkey. *Environment Monitoring and Assessment*, DOI 10.1007/s10661-008-0252-3.

Guo, Q., Kelly, M., Gong, P. and Liu, D. (2007). An object-based classification approach in mapping tree mortality using high spatial resolution imagery. *GIScience & Remote Sensing*, 44: 24 - 47.

Henning, J.G. (2005). Modeling Forest Canopy Distribution from Ground-based Laser Scanner Data, Ph.D. Dissertation, Virginia Polytechnic Institute and State University, Blacksburg, Virginia.

Herold, M., Liu, X. and Clarke, K. (2003). Spatial metrics and image texture for mapping urban land use. *Photogrammetric Engineering and Remote Sensing*, 69, 991–1001.

Hetzel, G., Leibe, B., Levi, P. and Schiele, B. (2001). 3D object recognition from range images using local feature histograms, Proceedings of the 2001 IEEE Computer Society Conference on Computer Vision and Pattern Recognition, 08–14 December, Kauai, Hawaii, IEEE Computer Society, Los Alamitos, California, pp. 394–399.

Hu, X., Tao, C.V. and Prenzel, B. (2005). Automatic segmentation of hig-resolution satellite imagery by integrating texture, intensity, and color features. *Photogrammetric Engineering and Remote Sensing*, 71, 1399-1406.

Huete, A. R. (1988). A soil-adjusted vegetation index (SAVI). *Remote Sensing of Environment*, 25, 295– 309.

Hyde, P., Dubayah, R., Peterson, B., Blair, J. B., Hofton, M. and Hunsaker, C. (2005). Mapping forest structure for wildlife habitat analysis using waveform lidar: Validation of montane ecosystems. *Remote Sensing of Environment*, 96(3–4), 427–437.

Hyde, P., Dubayah, R., Walker, W., Blair, B., Hofton, M. and Hunsaker, C. (2006). Mapping forest structure for wildlife habitat analysis using multi-sensor (LiDAR, SAR/InSAR, ETM+, Quickbird) synergy. *Remote Sensing of Environment*, 102, 63–73

Jarvis, P.G. and Leverenz, J.W. (1983). Productivity of temperate, deciduous and evergreen forests. In Physiological Plant Ecology IV. Encyclopedia of plant physiology. Vol. 12D. Lange , O.L., et al. (eds.) pp. 233-280. Springer-Verlag, New York.

Jensen, J.R. (2000). Remote sensing of the environment: an earth resource perspective, Prentice-Hall, Upper Saddle River, NJ.

Jordan, C. F. (1969). Derivation of leaf area index from quality of light on the forest floor. *Ecology*, 50, 663–666.

Kayitakire, F., Hamel, C., and Defourny, P. (2006). Retrieving forest structure variables based on image texture analysis and IKONOS-2 imagery. *Remote Sensing of Environment*, 102, 390–401.

Kaufman, Y. J. and Tanre, D. (1992). Atmospherically resistant vegetation index (ARVI) for EOS-MODIS, *IEEE Trans. Geosci. Remote Sens.* 30:261-270.

Liu, H.Q. and Huete, A. (1995). A feedback based modification of the NDVI to minimize canopy background and atmospheric noise. *IEEE Transactions on Geoscience and Remote Sensing*, 33, 457–465.

Loveland, T.R., Gallant, A.L. and Vogelmann, G.E. (2005). Perspectives on the use of land cover data for ecological investigations. In Issues and perspectives in landscape ecology. Eds. J. Wiens and M. Moss, Cambridge University Press.

Machado, J-L., & Reich, P. B. (1999). Evaluation of several measures of canopy openness as predictors of photosynthetic photon flux density in deeply shaded conifer-dominated forest understory. *Canadian Journal of Forest Research*, 29, 1438-1444.

Martinez Morales, R., Miura, T. and Idol, T. (2008). An assessment of Hawaiian dry forest condition with fine resolution remote sensing. *Forest Ecology and Management*, 255, 2524–2532.

Martinez Morales, R., Idol, T. and Friday, J.B. (2011). Assessment of *Acacia koa* forest health across environmental gradients in Hawai'i using fine resolution remote sensing and GIS, *Sensors*, 11, 5677-5694; doi:10.3390/s110605677.

Martinez Morales, R., Idol, T. and Chen, Q. (2012). Differentiation of *Acacia koa* forest stands across an elevation gradient in Hawai'i using fine resolution remotely sensed imagery. *International Journal of Remote Sensing*, 33(11), 3492-3511.

Plummer, S. E. (2000). Perspectives on combining ecological process models and remotely sensed data. *Ecological Modelling*, 129, 169–186.

Pu, R.L. and Gong, P. (2004). Wavelet transform applied to EO-1 hyperspectral data for forest LAI and crown closure mapping. *Remote Sensing of Environment*, 91(2), 212–224.

Pu, R., N. M. Kelly, Q. Chen, and Gong, P. (2008). Spectroscopic determination of health levels of Coast Live Oak (Quercus agrifolia) leaves. *Geocarto International*, 23(1), 3-20.

Qi, J., Chehbouni, A.I., Huete, A.R., Kerr, Y.H. and Sorooshian, S. (1994). A modified soil adjusted vegetation index (MSAVI). *Remote Sensing of Environment*, 48, 119-126.

Roberts, D.A., Green, R.O. and Adams, J.B. (1997). Temporal and spatial patterns in vegetation and atmospheric properties from AVIRIS. *Remote sensing of environment*, 62: 223 – 240.

Rouse, J. W. and Haas, R. H. (1973). Monitoring vegetation systems in the great plain with ERTS. Third ERTS Symposium, vol. 1 (pp. 309-317). Washington, DC: NASA.

Soudani, K., Francois, C., Maire, G. L., Dantec, V.L. and Dufrene, E. (2006). Comparative analysis of Ikonos, Spot, and ETM data for leaf area index estimation in temperate coniferous and deciduous forest stands. *Remote Sensing of Environment*, 102, 161–175.

Skole, D. and Tucker, C. (1993). Tropical deforestation and habitat fragmentation in the Amazon: Satellite data from 1978 to 1988. *Science*, 260, 1905–1909.

Tickle, P.K., Lee, A., Lucas, R.M., Austin, J. and Witte, C. (2006). Quantifying Australian forest floristics and structure using small footprint LiDAR and large scale aerial photography, *Forest Ecology and Management*, 223(1–3):379–394.

Treuhaft, R. N. and Siqueira, P. R. (2000). Vertical structure of vegetated land surfaces from interferometric and polarimetric radar. *Radio Science*, 35(1), 141–177.

Wang, L., Gong, P. and Biging, S. (2004). Individual tree-crown delineation and treetop detection in high-spatial-resolution aerial imagery. *Photogrammetric Engineering and Remote Sensing*, 70, 351–357.

Weishampel, J. F., Blair, J. B., Knox, R. G., Dubayah, R. and Clark, D. B. (2000). Volumetric lidar return patterns from an old-growth tropical rainforest canopy. *International Journal of Remote Sensing*, 21(2), 409–415.

Wulder, M., Hall, R., Coops, N. C. and Franklin, S. (2004). High spatial resolution remotely sensed data for ecosystem characterization. *BioScience*, 54(6), 511–521.

Van Dijk, A. and Bruijnzeel, L. A. (2001). Modelling rainfall interception by vegetation of variable density using an adapted analytical model: Part 1. Model description. *Journal of Hydrology*, 247, 230–238.

Xu, H. (2007). Extraction of Urban Built-up Land Features from Landsat Imagery Using a Thematicoriented Index Combination Technique. *Photogrammetric Engineering and Remote Sensing*, 73, 12:1381–1391.

7

Composition and Stand Structure of Tropical Moist Deciduous Forest of Similipal Biosphere Reserve, Orissa, India

R.K. Mishra[1], V.P. Upadhyay[2], P.K. Nayak[3],
S. Pattanaik[3] and R.C. Mohanty[3]
[1]Department of Wildlife and Biodiversity Conservation,
North Orissa University, Takatpur, Baripada
[2]Ministry of Environment and Forests, Eastern
Regional Office, Chandrasekharpur, Bhubaneswar
[3]Department of Botany, Utkal University, Vani Vihar, Bhubaneswar
India

1. Introduction

Tropical forests are highly productive, structurally complex, genetically rich renewable genetic resources (Roy et al., 2002). The tropical deforestation contributes to increase in atmospheric CO_2 and other gases affecting the climate and biodiversity. Though such type of forests occupy less than 7% of the land surface, there have the higher distinction of harbouring 50% of all plant and animal species (Mayers, 1992). The rate of forest loss due to deforestation as reported by Food and Agriculture Organistion [FAO, (2001)] is 15.2 million hectare per year (Data from 1990-2000). Assessment of the plant diversity of forest ecosystems is one of the fundamental goals of ecological research and is essential for providing information on ecosystem function and stability (World Conservation Monitoring Centre [WCMC], 1992; Tilman 2000; Townsend et. al., 2008). It has attracted attention of ecologists because of the growing awareness of its importance on the one hand and the massive depletion on the other (Singh, 2002; Lewis, 2009). Out of sixteen major forest types of India (Champion and Seth, 1968), tropical forests occupy 38 % of the total forest area in India (Dixit, 1997). However, in Orissa forest ecosystems cover about 37.34% of the State's geographical area and about 7.66% of country's forests. Large population of the state utilizes various components of the forests for both commercial and subsistence purposes. In the past few decades, heavy human pressure has reduced the forested area in the state resulting in degradation and fragmentation of historically contiguous landscapes posing threats to plant diversity (Murthy et al., 2007). It is now high time to conserve the plant diversity and has the task become a major concern for much of the society and for many governments and government agencies at all levels (Tripathi and Singh, 2009).

The Man and Biosphere Programme launched by United Nations Educational Scientific and Cultural Organisation ([UNESCO], 1971 as cited in Parker, 1984) aims at conserving the floral wealth in protected areas established by the Govt. of India in different states. Similipal Biosphere Reserve (SBR), a northern tropical moist deciduous type of forest (Champion and

Seth, 1968) situated in the Mayurbhanj district of Orissa has over the years, played important roles in maintaining the climate and livelihood of local communities (Srivastava and Singh, 1997; Rout et al., 2010). The SBR located in Eastern Ghat has distinctly dissimilar to the forests located in Western Ghats of India and Srilanka (Table-1). Table-2 provides a comparative account of floristic richness of some tropical forests. Genera like *Ficus, Diospyros, Syzygium, Symplocos, Dalbergia, Glochidion* are prominently represented in all these ecosystems as shown in Table-1.

Name of genera	Number of species		
	Western Ghats	Srilanka	Similipal
Goniothalamus (Annonaceae)	3	5	0
Garcinia (Clusiaceae)	9	6	0
Calophyllum (Clusiaceae)	3	9	0
Mesua (Clusiaceae)	1	3	1
Dipterocarpus (Dipterocarpaceae)	2	4	0
Hopea (Dipterocarpaceae)	8	4	0
Shorea (Dipterocarpaceae)	1	14	1
Stemonoporus (Dipterocarpaceae)	0	22	0
Pterospermum (Sterculiaceae)	6	0	1
Elaeocarpus (Elaeocarpaceae)	6	7	0
Ilex (Aquifoliaceae)	5	3	0
Euonymus (Celastraceae)	5	3	0
Holigarna (Anacardiaceae)	3	0	0
Semecarpus (Anacardiaceae)	2	10	1
Dalbergia (Leguminosae)	4	1	2
Humboldtia (Leguminosae)	5	1	0
Syzygium (Myrtaceae)	29	40	2
Memecylon (Melastomaceae)	9	25	1
Mastixia (Cornaceae)	1	3	0
Canthium (Rubiaceae)	5	4	0
Ixora (Rubiaceae)	6	4	1
Psychotria (Rubiaceae)	14	13	0
Lasianthus (Rubiaceae)	9	9	0
Vernonia (Asteraceae)	4	11	1
Ardisia (Myrsinaceae)	6	6	0
Palaquium (Sapotaceae)	2	9	0
Diospyros (Ebenaceae)	16	22	5
Symplocos (Symplocaceae)	21	12	2
Strobilanthus (Acanthaceae)	9	27	0
Myristica (Myristicaceae)	3	3	0
Cinnamomum (Lauraceae)	6	8	0
Actinodaphne (lauraceae)	5	9	0
Litsea (Lauraceae)	8	12	0
Cleistanthus (Euphorbiaceae)	2	5	1
Drypetes (Euphorbiaceae)	4	1	0
Glochidion (Euphorbiaceae)	10	9	2
Croton (Euphorbiaceae)	6	2	1
Agrostistachys (Euphorbiaceae)	2	3	0
Mallotus (Euphorbiaceae)	5	4	1
Macaranga (Euphorbiaceae)	2	3	1
Aporusa (Euphorbiaceae)	5	3	0
Ficus (Moraceae)	10	8	8

Table 1. Numbers of species in large woody plant genera confined to the Western Ghats, Srilanka and Similipal biosphere reserve (included under Eastern Ghats).

Forest locations	Area (ha)	Number of species	Number of genera	Number of families	Source of information
Jadkal forest	0.5	103	85	46	Vasanthraj et al., 2005
Lowland rain forest, Sabh, Malaysia	8.0	329	128	52	Campbell and Newbery, 1993
Low land dipterocarp forest, Danum Valley, Malaysia	8.0	511	164	59	Newbery et al., 1999
Keranga forest, Sarawak and Brunei	-	637	-	60	Newbery, 1991
Similipal, Orissa, India	3.6	266	204	76	Present study

Table 2. A comparative account of floristic richness of some tropical forest locations.

The National forest policy in India stipulates 33% of the total geographical area is to be under forest cover. Large area of fertile forest lands have been converted to other land uses to meet the demand of growing population. In addition opening of the close forests due to deforestation has resulted in increase in soil erosion, landslides, floods and loss of biodiversity and wildlife habitats. At the global level similar situation is reported from Brazil, Malyasia, Indonesia, Africa and Central American countries where loss of wildlife habitat ranges from 40-80% (Puri, 1995). The tropical dry forest of Coasta Rica (Heinrich and Hurka, 2004) is floristically very rich and diverse compared to the dry forests of Puerto Rico (Hare et al., 1997). Compared to other tropical dry deciduous forests of Eastern Ghats of India (Krishnannkutty et al., 2006) which are under various degrees of anthropogenic pressures, the SBR occupies strong ecological position in terms of species number and diversity. SBR is generally believed to be floristically rich, containing many varieties of plant life forms and medicinal plants as well (Saxena and Brahmam, 1989). Carefully compiled and up-to date information on diversity and distribution status of plant resources is however lacking. Though human-induced pressure, mainly through illegal chainsaw logging and access to non-timber forest products (NTFPs) is on the rise (Rout et al., 2009; Rout et al., 2010), a very few sporadic studies of SBR (Mishra et al., 2006, Mishra et al., 2008; Reddy et al., 2007) has so far been conducted to assess the plant diversity status. The conservation status of the biosphere reserve to be known attempting sustainable management, there should be need of proper documentation of diversity status of various plant life forms and their distribution patterns inside the reserve. Knowledge of floristic composition, structure and distribution of angiospermic plants of this biosphere reserve is critical in this direction.

2. Materials and methods

2.1 Study area

Similipal Biosphere Reserve (SBR) located between 21°28'- 22° 08' N latitude and 86°04' - 86°37' E longitude is situated in the Mayurbhanj district of Orissa stretching over an area of 5569 sq. km (Fig.1). The vast patch of forest covers of Similipal is one of the mega-biodiversity zones of the country with a rich population of flora and fauna. The elevation of valley peaks ranges from 80m to 869 m M.S.L. rolling with pockets of grassy meadows in between and traversed by a number of streams and waterfalls.

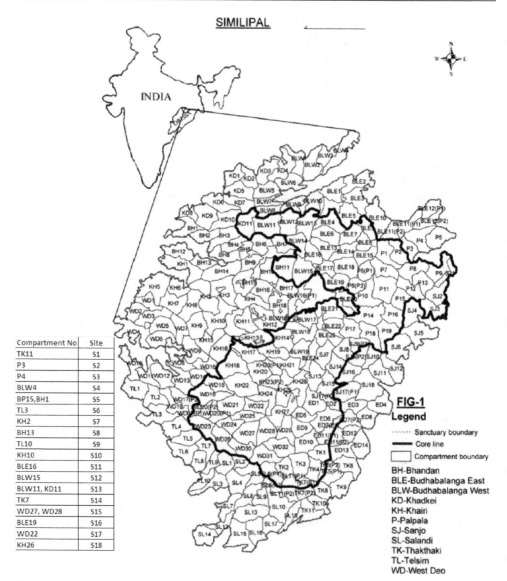

Compartment No	Site
TK11	S1
P3	S2
P4	S3
BLW4	S4
BP15, BH1	S5
TL3	S6
KH2	S7
BH13	S8
TL10	S9
KH10	S10
BLE16	S11
BLW15	S12
BLW11, KD11	S13
TK7	S14
WD27, WD28	S15
BLE19	S16
WD22	S17
KH26	S18

Fig. 1. Location map of sampling sites in Similipal biosphere reserve.

2.2 Climate

The climate of the reserve is influenced by a monsoon pattern of rainfall. Maximum rainfall occurs from mid June to October accounting for 75-80% of annual rainfall. In spite of high annual rainfall summer and winter are relatively dry generally with <10cm monthly rainfall (Mishra et al., 2006). The amount of average annual rainfall is not correlated with elevation and generally ranges from 28.11 to 344.96 cm. Summer is not unbearable, as the maximum temperature rarely goes above 40 ⁰C. Winter is severe and the temperature comes down to 4⁰C in some parts with frosts in valleys (Mishra et al., 2006). Spring is very pleasant. Because of good vegetation and a network of perennial streams Similipal is relatively moist throughout the year. Humidity of Similipal at 0600 hrs is around 40% and at 1800 hrs is around 81% to 93% (Srivastava and Singh, 1997).

2.3 Field methods (vegetation sampling and analysis)

To study the plant diversity status, 18 study sites were selected in East, West, North and South directions inside SBR (Fig.1).The vegetation analysis was conducted during 2005-2008 for all the six layers of the forest i.e. trees, climbers, shrubs, herbs, saplings and seedlings. The species were identified with flora guides (Saxena and Brahmam, 1994-1996; Haines, 1921-25). The tree layer was analyzed by sampling 20 quadrats of 10 m x 10 m size at each site. The size and number of samples were determined using the method of Kershaw (1973) and Mueller-Dombois and Ellenberg (1974). The abundance, density and frequency were calculated for the species. Importance Value Index (IVI) was determined as the sum of the relative frequency, relative density and relative dominance for tree layer only. Raunkiaer's frequency class (1934) analysis was used to assess the rarity or commonness of the tree species (Hewit and Kellman, 2002). In this classification the percentage frequency of the species was classed as A, B, C, D and E; where A represents rare (0–20%), B is low frequency (20–40%), C is intermediate frequency (40–60%), D is moderately high frequency (60–80%) and E is high frequency or common (80–100%). With this classification, the expected distribution of the species is A>B>C≤ ≥D<E. The distribution pattern of different species was studied using the ratio of abundance to frequency (Whitford, 1949). Trees were ≥ 30cm cbh (circumference at breast height), saplings were 10-30 cm cbh and seedlings were <10cm cbh (Knight, 1975). The shrub and herb layers were analyzed by randomly placing 20 quadrats of 5m x 5m size and 1m x 1m size, respectively at each site during the post monsoon season. The diversity index at each site was computed by using Shannon- Wiener information function (Shannon-Wiener, 1963) and concentration of dominance by Simpson's index (Simpson, 1949), evenness and richness index following Pielou (1975) and Margalef (1958) (as cited in Tripathi and Singh, 2009), respectively. The presence of climbers on trees affects their growth and development. They have been noted to suppress natural regeneration and delay forest recovery (Babweteera et. al., 2001). The presence or absence of climbers on the trees was scored on a 5-point scale (Alder and Synnott, 1992) whereby 1, 2, 3, 4 and 5 represented trees that were: having bore climbers; trees over grown with climbers; climbers on the stem only; climbers in the crown only and climbers both on the stem and crown, respectively.

3. Results

3.1 Floristic composition and occurrence

A total of 266 species belonging to 204 genera and 76 families were recorded from the study area, out of which 117 were tree species, 17 climber, 31 shrub and 101 herb species. Thus only approximately 24.72% of the estimated flora of Similipal (Saxena and Brahmam, 1989) was covered in the study (Table-3, 4, 5 and 6). A majority of the families were represented by only two or less species. The most common families were Euphorbiaceae and Rubiaceae, each represented by 19 species; followed by Fabaceae (15 species), Mimosaceae = Acanthaceae (12 species each), Asteraceae (11 species), Cyperaceae= Moraceae= Caesalpinaceae = Combretaceae (9 species each), Malvaceae = Melastomataceae = Rutaceae = Poaceae (7 species each), etc. The average number of species per hectare was 74. The number of species per genus was 1.3 and that per family was 3.5. Species in various groups of plant life forms had a wide range of occurrence, ranging in frequency from 5- 72% in herbs, 5-94% in shrubs, and 5- 100% in case of trees, climbers, saplings and seedlings (Table-3, 4, 5, 6, 7 and 8).

Name of the family	Name of the plant species	Density (Plants/ha)	Basal Area (m^2/ha)	Frequency (%)	IVI	A /F
Rubiaceae	*Adina cordifolia* (Roxb.) Hook. f.ex. Brandis	11.94	1.44	77.78	6.54	0.04
Rutaceae	*Aegle marmelos* (L.) Corr.	4.72	0.353	44.44	2.81	0.05
Mimosaceae	*Albizia marginata* (Lam.) Merr.	3.33	0.37	22.22	1.80	0.14
Combretaceae	*Anogeissus latifolia* (Roxb. ex DC.) Wall ex. Guill	45.28	3.36	77.78	13.43	0.15
Bombacaceae	*Bombax ceiba* L.	10.83	1.53	55.56	5.64	0.07
Euphorbiaceae	*Bridelia retusa* (L.) Spreng.	4.72	0.45	38.89	2.73	0.06
Anacardiaceae	*Buchanania lanzan* Spreng.	19.44	0.81	66.67	6.16	0.09
Lecythidaceae	*Careya arborea* Roxb.	6.67	0.36	44.44	3.07	0.07
Flacourtiaceae	*Casearia graveolens* Dalz.	5.28	0.12	55.56	2.98	0.03
Caesalpiniaceae	*Cassia fistula* L.	6.67	0.12	44.44	3.06	0.07
Euphorbiaceae	*Cleistanthus collinus* (Roxb.) Benth. ex Hook.f.	3.33	0.06	11.11	0.94	0.54
Euphorbiaceae	*Croton roxburghii* Balak	6.11	0.21	22.22	1.92	0.25
Mimosaceae	*Dalbergia latifolia* Roxb.	2.78	0.21	22.22	1.41	0.11
Fabaceae	*Desmodium oojeinesis* (Roxb.) Ohashi	5.00	0.58	27.78	2.53	0.13
Dilleniaceae	*Dillenia pentagyna* Roxb.	29.17	2.53	77.78	10.24	0.10
Ebenaceae	*Diospyros embryopteris* Pers.	2.78	0.17	22.22	1.44	0.11
Ebenaceae	*Diospyros melanoxylon* Roxb.	8.61	0.66	44.44	3.73	0.09
Ebenaceae	*Diospyros montana* Roxb.	2.78	0.07	22.22	1.34	0.11
Burseraceae	*Garuga pinata* Roxb.	2.78	0.16	27.78	1.65	0.07
Simaroubaceae	*Gmelina arborea* Roxb.	5.28	0.44	50.00	3.21	0.04
Apocynaceae	*Holarrhena antidysenterica* Wall.ex A.DC.	2.78	0.07	22.22	1.31	0.11
Malvaceae	*Kydia calycina* Roxb.	5.56	0.22	38.89	2.51	0.07
Lythraceae	*Lagerstroemia parviflora* Roxb.	6.39	0.39	33.33	2.65	0.12
Anacardiaceae	*Lannea corromandelica* (Houtt.) Merr.	4.72	0.84	27.78	2.85	0.12

Name of the family	Name of the plant species	Density (Plants/ha)	Basal Area (m²/ha)	Frequency (%)	IVI	A /F
Sapotaceae	Madhuca latifolia Gmel.	12.78	1.074	44.44	4.84	0.13
Anacardiaceae	Mangifera indica L.	3.61	1.08	38.89	3.48	0.05
Magnoliaceae	Michelia champaca L.	7.78	0.98	11.11	2.78	1.26
Oleaceae	Nyctanthes arbor- tristis L.	3.89	0.15	27.78	1.78	0.10
Ochnaceae	Ochna obtusata DC.	5.83	0.35	16.67	1.87	0.42
Fabaceae	Pterocarpus marsupium Roxb.	12.22	1.24	55.56	5.43	0.08
Euphorbiaceae	Phyllanthus emblica L.	4.17	0.2	27.78	1.88	0.11
Burseraceae	Protium serratum (Wall. ex Colebr.) Engl.	32.22	1.99	83.33	10.08	0.09
Mimosaceae	Samanea saman (Jacq.) Merr.	2.78	0.13	3.13	1.40	0.80
Sapindaceae	Schleichera oleosa (Lour.) Oken	13.33	1.48	33.33	5.05	0.24
Euphorbiaceae	Securinega virosa (Roxb. ex Willd.) Baill	11.11	1.02	33.33	4.13	0.20
Dipterocarpaceae	Shorea robusta Gaertn.f.	284.17	27.73	100.00	77.67	0.57
Myrtaceae	Syzygium cumini (L.) Skeels	23.06	2.03	83.33	10.19	0.07
Myrtaceae	Syzygium cerasoides (Roxb.)Chatt. & Kanjlal	18.06	1.27	66.67	6.64	0.08
Combretaceae	Terminalia alata Heyne ex Roth.	50.28	4.37	94.44	16.13	0.11
Combretaceae	Terminalia bellirica (Gaertn.) Roxb.	6.67	0.44	44.44	3.49	0.07
Combretaceae	Terminalia chebula Retz.	5.56	0.78	61.11	4.29	0.03
Verbenaceae	Vitex leucoxylon (L.f.)	5.56	0.21	38.89	2.50	0.07
Rubiaceae	Wendlandia tinctoria (Roxb.) DC.	3.61	0.21	27.78	1.82	0.09
Mimosaceae	Xylia xylocarpa (Roxb.) Taub.	2.78	0.2	16.67	1.27	0.20
Rhamnaceae	Ziziphus mauritiana Lam.	0.28	0.04	5.56	0.31	0.18
Rhamnaceae	Ziziphus rugosa Lam.	2.78	0.05	27.78	1.48	0.07
Total		793.67	71.043	-	299.75	-

Table 3. Families, species, density, basal area, frequency, distribution pattern and Importance Value Index (IVI) of trees in Similipal biosphere reserve.

Name of the family	Name of the plant species	Density (Individuals/ha)	Frequency %	Abundance	A/F
Euphorbiaceae	*Antidesma ghaesembila* Gaertn.	97.78	94.44	5.18	0.05
Myrsinaceae	*Ardisia solanacea* Roxb.	45.56	55.56	4.10	0.07
Violaceae	*Bixa orellana* L.	17.78	22.22	4.00	0.18
Rubiaceae	*Catunaregam spinosa* (Thunb.) Tirveng.	12.22	16.67	3.67	0.22
Meliaceae	*Cipadessa baccifera* (Roxb.) Miq.	3.33	5.56	3.00	0.54
Rutaceae	*Citrus medica* L.	48.89	44.44	5.50	0.12
Rutaceae	*Clausena excavata* Burm. f.	3.33	5.56	3.00	0.54
Verbenaceae	*Clerodendrum serratum* (L.) Moon	32.22	22.22	7.25	0.33
Euphorbiaceae	*Croton caudatus* Geisel.	20.00	27.78	3.60	0.13
Rubiaceae	*Gardenia resinifera* Roth	17.78	27.78	3.20	0.12
Euphorbiaceae	*Glochidion* sp.	15.56	22.22	3.50	0.16
Lamiaceae	*Gomphostemma parviflorum* Wall. ex Benth.	21.11	27.78	3.80	0.14
Tiliaceae	*Grewia hirsuta* Vahl.	22.22	27.78	4.00	0.14
Sterculiaceae	*Helicteres isora* L.	17.78	22.22	4.00	0.18
Euphorbiaceae	*Homonoia riparia* Lour.	22.22	44.44	2.50	0.06
Hypericaceae	*Hypericum gaitii* Haines	54.44	22.22	12.25	0.55
Rubiaceae	*Hyptianthera sticta* (Wild.) Wight & Arn.	144.44	66.67	10.83	0.16
Fabaceae	*Indigofera cassioides* Rottel ex DC.	125.56	50.00	12.56	0.25
Oleaceae	*Jasminum arborescens* Roxb.	10.00	27.78	1.80	0.06
Verbenaceae	*Lantana camara* L.	287.78	61.11	23.55	0.39
Vitaceae	*Leea asiatica* (L.) Ridsdale	15.56	22.22	3.50	0.16
Vitaceae	*Leea indica* (Burm. f.) Merr.	34.44	16.67	10.33	0.62
Melastomataceae	*Melastoma malabathricum* L.	122.22	50.00	12.22	0.24
Rubiaceae	*Pavetta tomentosa* Roxb. Ex Sm.	22.22	27.78	4.00	0.14
Lamiaceae	*Pogostemon benghalensis* (Burm. f.) Kuntze	341.11	55.56	30.70	0.55
Fabaceae	*Sesbania bispinosa* (Jacq.) W. F. Wight	18.89	22.22	4.25	0.19
Malvaceae	*Urena lobata* L.	46.67	38.89	6.00	0.15
Asteraceae	*Vernonia anthelmintica* (L.) Willd.	30.00	27.78	5.40	0.19
Lythraceae	*Woodfordia fruticosa* (L.) Kurz	188.89	77.78	12.14	0.16
Rubiaceae	*Gardenia gummifera* L. f.	31.11	22.22	7.00	0.32
Fabaceae	*Flemingia chappar* Buch. – Ham. ex Benth.	73.33	50.00	7.33	0.15
Total		1944.44	-	-	-

Table 4. Families, species, density, frequency, abundance and distribution pattern of shrub layer in Similipal biosphere reserve.

Name of the family	Name of the plant species	Density (Individuals/ha)	Frequency %	Abundance	A/F
Malvaceae	*Abutilon indicum* (L.) Sweet	1944.44	16.67	23.33	1.40
Asteraceae	*Ageratum conyzoides* L.	1527.78	38.89	7.86	0.20
Acanthaceae	*Andrographis paniculata* (Bum.f.) Wall.ex. Nees	833.33	11.11	15.00	1.35
Commelinaceae	*Aneilema ovalifolium* (Wight) Hook.f.ex.	1472.22	11.11	26.50	2.39
Scrophulariaceae	*Bacopa monieri* (L.) Pennell.	2000.00	5.56	72.00	12.96
Capparaceae	*Cleome viscosa* L.	1277.78	11.11	23.00	2.07
Commelinaceae	*Commelina benghalensis* L.	1083.33	11.11	19.50	1.76
Commelinaceae	*Commelina palludosa* Bl.	583.33	5.56	21.00	3.78
Commelinaceae	*Commelina* sp.	611.11	11.11	11.00	0.99
Zingiberaceae	*Costus speciosus* (Koeing) Sm.	1555.56	5.56	56.00	10.08
Hypoxidaceae	*Curculigo orchoides* Gaertn.	6638.89	72.22	18.38	0.25
Zingiberaceae	*Curcuma amada* Roxb.	14138.89	55.56	50.90	0.92
Zingiberaceae	*Curcuma aromaticum* Salisb.	2111.11	27.78	15.20	0.55
Cyperaceae	*Cyperus rotundus* L.	7611.11	11.11	137.00	12.33
Cyperaceae	*Cyperus* sp.	1527.78	11.11	27.50	2.48
Fabaceae	*Desmodium trifolium* (L.) DC.	35888.89	27.78	258.40	9.30
Acanthaceae	*Dicliptera bleupleuroides* Mees.	861.11	11.11	15.50	1.40
Poaceae	*Eragrostis cilliata* (Roxb.) Nees	1805.56	44.44	8.13	0.18
Acanthaceae	*Eranthemum purpurascens* Nees	916.67	22.22	8.25	0.37
Convolvulaceae	*Evolvulus alsinoides* L.	21333.33	27.78	153.60	5.53
Convolvulaceae	*Evolvulus numularis* (L.) L.	14694.44	11.11	264.50	23.81
Cyperaceae	*Fimbristylis aestivalis* (Retz.) Vahl.	3138.89	11.11	56.50	5.09
Rubiaceae	*Knoxia sumatrensis* (Retz.) DC.	3861.11	11.11	69.50	6.26
Lobeliaceae	*Lobelia alsinoides* Lam.	888.89	11.11	16.00	1.44
Cyperaceae	*Mariscos* sp.	1500.00	11.11	27.00	2.43
Sterculiaceae	*Melotia curcurifolia* L.	777.78	5.56	28.00	5.04
Mimosaceae	*Mimosa pudica* L.	611.11	5.56	22.00	3.96
Orobanchaceae	*Orthosiphon rubicundus* (D.Don) Benn.	805.56	11.11	14.50	1.31
Urticaceae	*Pozoulzia pentandra* (Roxb.) Benn.	1111.11	16.67	13.33	0.80
Amaryllidaceae	*Pancratium trifolium* Roxb.	1722.22	22.22	15.50	0.70
Acanthaceae	*Phlogacanthus* sp.	611.11	5.56	22.00	3.96
Arecaceae	*Phoenix acaulis* Buch-Ham.ex Roxb.	888.89	22.22	8.00	0.36
Arecaceae	*Phoenix* sp.	666.67	11.11	12.00	1.08
Euphorbiaceae	*Phyllanthus fraternus*	3416.67	38.89	17.57	0.45
Acanthaceae	*Rungia pectinata* (L.) Nees ex DC.	1750.00	27.78	12.60	0.45
Amaranthaceae	*Celosia argentia* (L.)	583.33	5.56	21.00	3.78
Fabaceae	*Shuteria involucrata* (Wall.) Wt. & Arn.	583.33	27.78	4.20	0.15
Malvaceae	*Sida cordifolia* L.	1111.11	11.11	20.00	1.80

Name of the family	Name of the plant species	Density (Individuals/ha)	Frequency %	Abundance	A/F
Rubiaceae	*Spermococe pusilla* Wall.	1527.78	5.56	55.00	9.90
Acanthaceae	*Strobilanthus auriculatus* Nees	888.89	22.22	8.00	0.36
Fabaceae	*Uraria picta* (Jacq.) Desv. Ex DC.	1138.89	16.67	13.67	0.82
Fabaceae	*Zonia dicola*	1166.67	11.11	21.00	1.89
Rubiaceae	*Hediotys verticillata* (L.) Lam.	861.11	27.78	6.20	0.22
Orchidaceae	*Eulophia nuda* Lindi.	1027.78	33.33	6.17	0.19
Acanthaceae	*Barleria srigosa* (Wild.)	611.11	38.89	3.14	0.08
Cyperaceae	*Cyperus triceps* Endl.	861.11	27.78	6.20	0.22
Others	Others	14083.3	5.56 – 27.78	2 - 18	0.06 – 3.24
Total		1,69,500	-	-	-

Table 5. Families, species, density, frequency, abundance and distribution pattern of herb layer in Similipal biosphere reserve.

Name of the family	Name of the plant species	Density (Individuals/ha)	Frequency %	Abundance	A/F
Liliaceae	*Asparagus racemosus* Willd	9.44	38.89	4.86	0.12
Caesalpinaceae	*Bauhinia vahlii* wight. &Arn	13.33	100.00	2.67	0.03
Fabaceae	*Butea superba* Roxb	3.33	33.33	2.00	0.06
Combretaceae	*Calycopteris floribunda* Lam	3.61	38.89	1.86	0.05
Combretaceae	*Combretum roxburghii* Spreng	5.28	55.56	1.90	0.03
Dioscoreaceae	*Dioscorea bulbifera* L.	8.06	38.89	4.14	0.11
Mimosaceae	*Entada rheedii* Spreng	1.39	16.67	1.67	0.10
Asclepiadaceae	*Hemidesmus indicus* (L.) R.Br.	6.39	44.44	2.88	0.06
Fabaceae	*Millettia extensa* (Benth.) Baker	8.33	27.78	6.00	0.22
Asclepiadaceae	*Pergularia daemia* (Forssk.) Chiov.	13.06	44.44	5.88	0.13
Liliaceae	*Smilax macrophylla* Roxb	8.89	38.89	4.57	0.12
Liliaceae	*Smilax prolifera* Wall ex Roxb.	0.56	5.56	2.00	0.36
Apocynaceae	*Aganosma caryophyllata* (Roxb.ex sims)G.Don	1.94	16.67	2.33	0.14
Euphorbiaceae	*Bridelia stipularis* Bl.	0.83	5.56	3.00	0.54
Lygodiaceae	*Lygodium flexicosum* (L.) Sw	1.39	11.11	2.50	0.23
Araceae	*Pothos scandens* L.	1.94	16.67	2.33	0.14
Oleaceae	*Jasminum flexile* Vahl	0.83	11.11	1.50	0.14
Total		88.6	-	-	-

Table 6. Families, species, density, frequency, abundance and distribution pattern of climber layer in Similipal biosphere reserve.

Name of the family	Name of the plant species	Density (Individuals/ha)	Frequency %	Abundance	A/F
Rubiaceae	*Adina cordifolia* (Roxb.) Hook. f.ex. Brandis	18.89	44.44	2.13	0.05
Rutaceae	*Aegle marmelos* (L.) Corr.	6.67	11.11	3.00	0.27
Mimosaceae	*Albizia marginata* (Lam.) Merr.	12.22	11.11	5.50	0.50
Combretaceae	*Anogeissus latifolia* (Roxb. ex DC.) Wall ex. Guill	85.56	88.89	4.81	0.05
Barringtoniaceae	*Baringtonia acutangula* (L.) Gaertn.	11.11	16.67	3.33	0.20
Caesalpinaceae	*Bauhinia variegata* L.	8.89	16.67	2.67	0.16
Bombacaceae	*Bombax ceiba* L.	6.67	16.67	2.00	0.12
Anacardiaceae	*Buchanania lanzan* Spreng	58.89	55.56	5.30	0.10
Lecythidaceae	*Careya arborea* Roxb.	20.00	44.44	2.25	0.05
Flacourtiaceae	*Casearia graveolens* Dalz.	64.44	83.33	3.87	0.05
Caesalpinaceae	*Cassia fistula* L.	27.78	50.00	2.78	0.06
Cochlospermaceae	*Chochlospermum gossypium* DC.	7.78	5.56	7.00	1.26
Meliaceae	*Cipadessa baccifera* (Roxb.) Miq.	12.22	22.22	2.75	0.12
Euphorbiaceae	*Cleistanthus collinus* (Roxb.) Benth.-ex Hook. f.	18.89	16.67	5.67	0.34
Euphorbiaceae	*Croton roxburghii* Balak	14.44	27.78	2.60	0.09
Fabaceae	*Desmodium oojeinesis* (Roxb.) Ohashi	5.56	11.11	2.50	0.23
Dilleniaceae	*Dillenia pentagyna* Roxb.	50.00	61.11	4.09	0.07
Ebenaceae	*Diospyros malabarica* (Desr.) Kostel.	8.89	11.11	4.00	0.36
Ebenaceae	*Diospyros melanoxylon* Roxb.	21.11	33.33	3.17	0.10
Euphorbiaceae	*Glochidion lanceolarium* (Roxb.) Dalz.Glochidion	8.89	11.11	4.00	0.36
Simaroubaceae	*Gmelina arborea* Roxb.	8.89	22.22	2.00	0.09
Sterculiaceae	*Helicteres isora* L.	10.00	22.22	2.25	0.10
Apocynaceae	*Holarrhena antidysentrica* Wall.ex A.DC.	7.78	16.67	2.33	0.14
Flacourtiaceae	*Homalium nepalens* Benth.	40.00	61.11	3.27	0.05
Malvaceae	*Kydia calycina* Roxb.	6.67	11.11	3.00	0.27
Anacardiaceae	*Nothopegia heyneana* (Hook. f.)	5.56	5.56	5.00	0.90
Lythraceae	*Lagerstroemia parviflora*	17.78	27.78	3.20	0.12

Name of the family	Name of the plant species	Density (Individuals/ha)	Frequency %	Abundance	A/F
	Roxb.				
Sapotaceae	*Madhuca latifolia* Gmel.	13.33	27.78	2.40	0.09
Annonaceae	*Miliusa velutina* (Dunal) Hook.f . & Thomas.	7.78	16.67	2.33	0.14
Rubiaceae	*Mitragyna parviflora* (Roxb.) Korth.	5.56	11.11	2.50	0.23
Oleaceae	*Nyctanthes arber-tristis* L.	36.67	72.22	2.54	0.04
Euphorbiaceae	*Phyllanthus emblica* L.	33.33	61.11	2.73	0.04
Burseraceae	*Protium serratum* (Wall. ex Colebr.) Engl.	35.56	55.56	3.20	0.06
Sapindaceae	*Schleichera oleosa* (Lour.) Oken	45.56	66.67	3.42	0.05
Euphorbiaceae	*Securinega virosa* (Roxb. ex Willd.) Baill	38.89	50.00	3.89	0.08
Dipterocarpaceae	*Shorea robusta* Gaertn.f.	365.56	100.00	18.28	0.18
Sterculiaceae	*Sterculia urens* Roxb.	17.78	16.67	5.33	0.32
Bignoniaceae	*Sterospermum suaveolens* (Roxb.)DC.	13.33	16.67	4.00	0.24
Myrtaceae	*Syzygium cerasoides* (Roxb.)Chatt. & Kanjlal	18.89	33.33	2.83	0.09
Myrtaceae	*Syzigium cumini* (L.) Skeels	30.00	55.56	2.70	0.05
Combretaceae	*Terminalia alta* Heyne ex Roth.	74.44	88.89	4.19	0.05
Combretaceae	*Terminalia bellirica* (Gaertn.) Roxb.	18.89	27.78	3.40	0.12
Combretaceae	*Terminalia chebula* Retz.	11.11	33.33	1.67	0.05
Rubiaceae	*Wendlandia tinctoria* (Roxb.) DC.	20.00	33.33	3.00	0.09
Mimosaceae	*Xylia xylocarpa* (Roxb.) Taub.	15.56	27.78	2.80	0.10
Rhamnaceae	*Ziziphus rugosa* Lam.	7.78	22.22	1.75	0.08
Others		148.74	5.56-16.67	1.00-4.00	0.08-0.72
Total		1524.34	-	-	-

Table 7. Families, species, density, frequency, abundance and distribution pattern of sapling layer in Similipal biosphere reserve.

Name of the family	Name of the plant species	Density (Individuals/ha)	Frequency %	Abundance	A/F
Dipterocarpaceae	*Shorea robusta* Gaertn.f.	27777.78	55.56	100.00	0.56
Euphorbiaceae	*Croton roxburghii* Balak	24250.00	145.50	33.33	4.37
Combretaceae	*Terminalia alata* Heyne ex Roth.	3416.67	9.46	72.22	0.13
Anacardiaceae	*Buchanania lanzan* Spreng.	4944.44	14.83	66.67	0.22
Ebenaceae	*Diospyros melanoxylon* Roxb.	5888.89	26.50	44.44	0.60
Euphorbiaceae	*Cleistanthus collinus* (Roxb.) Benth. Ex Hook.f.	4611.11	23.71	38.89	0.61
Sterculiaceae	*Sterculia urens* Roxb.	1861.11	22.33	16.67	1.34
Flacourtiaceae	*Homalium nepalens* Benth.	1611.11	8.29	38.89	0.21
Euphorbiaceae	*Phyllanthus emblica* L.	2138.89	12.83	33.33	0.39
Combretaceae	*Anogeissus latifolia* (Roxb. Ex DC.) Wall ex. Guill	1500.00	13.50	22.22	0.61
Rutaceae	*Aegle marmelos* (L.) Corr.	972.22	7.00	27.78	0.25
Mimosaceae	*Dalbergia sisoo* Roxb.	500.00	18.00	5.56	3.24
Sapindaceae	*Schleichera oleosa* (Lour.) Oken	805.56	4.14	38.89	0.11
Simaroubaceae	*Ailanthus* sp. Roxb.	805.56	29.00	5.56	5.22
Apocynaceae	*Hollarhaena antidysentrica* Wall.ex A.DC.	1361.11	7.00	38.89	0.18
Oleaceae	*Nyctanthes arbortristis* L.	1527.78	18.33	16.67	1.10
Rhamnaceae	*Ziziphus rugosa* Lam.	916.67	5.50	33.33	0.17
Euphorbiaceae	*Bridelia retusa* (L.) Spreng.	1277.78	11.50	22.22	0.52
Rubiaceae	*Ixora* sp.	722.22	26.00	5.56	4.68
Rubiaceae	*Gardenia gummifera* .L.f.	361.11	4.33	16.67	0.26
Fabaceae	*Pterocarpus marsupium* Roxb.	916.67	5.50	33.33	0.17
Fabaceae	*Desmodium oojeinensis* (Roxb.) Ohashi	1194.44	10.75	22.22	0.48
Myrtaceae	*Syzygium cumini* (L.) Skeels	2472.22	7.42	66.67	0.11
Rubiaceae	*Wendlandia* sp.	2000.00	72.00	5.56	12.96
Sterculiaceae	*Helicteres isora*. L.	972.22	8.75	22.22	0.39
Mimosaceae	*Albizia odoratissima*.	1222.22	22.00	11.11	1.98

Name of the family	Name of the plant species	Density (Individuals/ha)	Frequency %	Abundance	A/F
	(L.f.) Benth.				
Bignoniaceae	Sterospermum suaveolens (Roxb.) DC.	638.89	7.67	16.67	0.46
Combretaceae	Terminalia chebula Retz.	861.11	5.17	33.33	0.16
Meliaceae	Trichilia connaroides (Wight & Arn.) Bentv.	638.89	23.00	5.56	4.14
Euphorbiaceae	Securinega virosa (Roxb. Ex Wild.) Baill	500.00	6.00	16.67	0.36
Flacourtiaceae	Casearia graveolens Dalz.	1750.00	6.30	55.56	0.11
Dilleniaceae	Dillenia pentagyna Roxb.	2194.44	11.29	38.89	0.29
Mimosaceae	Xylia xylocarpa (Roxb.) DC.	1583.33	19.00	16.67	1.14
Mimosaceae	Dalbergia latifolia Roxb.	805.56	29.00	5.56	5.22
Burseraceae	Protium serratum (Wall.ex Colebr.) Engl.	694.44	3.57	38.89	0.09
Mimosaceae	Albizia marginata (Lam.) Merr.	861.11	15.50	11.11	1.40
Rubiaceae	Adina cordifolia (Roxb.) Hook. F.ex. Brandis	361.11	6.50	11.11	0.59
Anacardiaceae	Mangifera indica L.	527.78	4.75	22.22	0.21
Sapotaceae	Madhuca latifolia Gmel.	583.33	10.50	11.11	0.95
Lecythidaceae	Careya arborea Roxb.	527.78	4.75	22.22	0.21
Euphorbiaceae	Securinega virosa (Roxb. Ex Wild.) Baill	583.33	21.00	5.56	3.78
Simaroubaceae	Gmelina arborea Roxb.	416.67	5.00	16.67	0.30
Mimosaceae	Albizia procera (Roxb.) Benth	388.89	14.00	5.56	2.52
Bombacaceae	Bombax ceiba L.	388.89	7.00	11.11	0.63
Euphorbiaceae	Mallotus phillipensis (Lam.) Muell.-Arg.	416.67	15.00	5.56	2.70
Lythraceae	Lagerostroemia parviflora Roxb.	333.33	12.00	5.56	2.16
Rubiaceae	Wendlandia tinctoria (Roxb.) DC.	416.67	15.00	5.56	2.70
Meliaceae	Cipadessa baccifera(Roxb.) Miq	333.33	12.00	5.56	2.16
		1444.44	1.00-9.00	5.56-11.11	0.18-1.62
Total		1,13,416.7	-	-	-

Table 8. Families, species, density, frequency, abundance and distribution pattern of seedling layer in Similipal biosphere reserve.

In most of the plant life forms there were a high number of species that occurred only once. The distribution of the species into Raunkiaer's frequency classes showed that most of the species encountered were rare and very few species were in intermediate and high or common frequency class (Table- 9). Except climbers all other groups of plant life forms do not follow the expected A>B>C ≥ ≤ D<E frequency distribution proposed by Raunkiaer (1934) as the number of species in frequency class D is higher than frequency class E.

Frequency Class	Code	Number of species in vegetation layers						
		Tree	Climber	Shrub	Herb	sapling	seedling	Remark
0-20	A	71 (61)	7 (41)	4 (13)	78 (77)	79 (73)	34 (59)	Rare
21-40	B	27(23)	6 (35)	16 (52)	20 (20)	13 (12)	18 (31)	Low
41-60	C	9 (08)	3 (18)	7 (22)	2 (02)	7 (06)	2 (03)	Intermediate frequency
61-80	D	6 (05)	0 (0)	3 (10)	1 (01)	5 (05)	3 (05)	Moderately high frequency
81-100	E	4 (03)	1 (06)	1 (03)	0 (0)	4 (04)	1 (02)	High frequency (common)

Table 9. Distribution of vegetation layers according to Raunkiaer's classification scheme (Values in parentheses indicate % of species).

3.2 Ecological importance of species

Importance value Index (IVI) is the measurement of ecological amplitude of species (Ludwig and Reynolds, 1988) suggesting the ability of a species to establish over an array of habitats. However, there is no single perfect way of assessing the ecological amplitude of a species. The abundance of a species can be represented by several measures such as relative density, relative frequency and Importance Value Index (IVI). Though frequency and density values are suitable for herbs and shrubs (Airi et al., 2000), IVI is an important information for tree species. On the basis of IVI, *Shorea robusta* was found as the dominant species in the SBR having IVI of 77.67 followed by *Terminalia alata* (16.13) and *Anogeissus latifolia* (13.43). *Wendlandia* sp. had IVI of 0.25 and was considered as the rare species of the reserve. All other tree species showed intermediate range of IVI (Table-3).

3.3 Distribution pattern

The distribution pattern of trees, shrubs, climbers, herbs, saplings and seedlings of the reserve is shown in Table-10. Odum (1971) stated that under natural conditions, a clumped distribution of plants is normal. A higher percentage of random and regular distribution reflects the greater magnitude of disturbance` such as grazing and lopping in natural forest stands. Most of the species of all the vegetational layers of the reserve showed generally clumped type of distribution in the present study. Regular distribution pattern is completely lacking in all the vegetation layers. Both in herb and seedling layers not a single species showed random distribution pattern (Table-10).

Plant group	Number of species in distribution pattern categories			
	Regular	Random	Contiguous	Total number of species
Tree	0	08	109	117
Climber	0	03	14	17
Shrub	0	01	30	31
Herb	0	0	101	101
Sapling	0	06	102	108
Seedling	0	0	58	58

Table 10. Distribution pattern of vegetation layers of Similipal biosphere reserve.

3.4 Distribution of climbers

Out of 794 number of trees per hectare 110 number of trees per hectare affected by 40% bore climbers, 10 % were overgrown with climbers while 15% had climbers restricted to the main stem, 12% had climbers in the crown only and 23% had climbers both the stem and in the crown (Figure. 2).

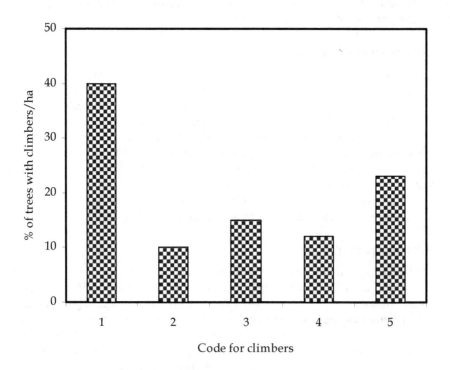

Fig. 2. Percentage distribution of climbers in Similipal biosphere reserve.

3.5 Stand structure

Species wise density of individuals having ≥30cm girth of the reserve ranged from less than one plant per hectare to 284 plants/ha and the total density of the reserve was 794 plants/ha. Maximum density (per hectare of individuals of ≥30cm) was recorded for *Shorea robusta* (284) followed by *Terminalia alata* (50), *Anogeissus latifolia* (45), *Protium serratum* (32) and *Dillenia pentagyna* (29). Density was observed less than or equal to one for many species like *Antidesma acidum, Artocarpus lacuccha, Butea monosperma,, Casearia elliptica, Chionanthus intermedicus, Cochlospermum religiosum, Euonymus glaber*, etc. Some other species showed intermediate range of density per hectare. The densities of climbers in comparison to other vegetation layers of the reserve was too low. However the densities of herbs, seedlings and sapling layers were quite high in comparison to other vegetational layers. Unlike tree layer in herb, shrub and climber layers very few species showed lowest density. *Exacum bicolor* in herb layer, *Cipadesa baccifera* and *Clausena excavata* in shrub layer and, *Jasminum flexile* and *Bridelia stipularis* in climber layers values showed minimum value of density (Table-4, 5, 6, 7 and 8). Total basal area of trees of the reserve was 71.05 m²/ha in which maximum was experienced by *Shorea robusta*. *Shorea robusta* contributed maximum of 39% to the basal area followed by *Terminalia alata* (6.15%), *Anogeissus latifolia* (4.73%) and *Dillenia pentagyna* (3.57%). The total contribution that resulted from this associated combination of *Shorea-Terminalia-Anogeissus-Dillenia* was 53.45%. A few families contributed most to the total basal area. These included Dipterocarpaceae (39%), Combretaceae (13 %), Myrtaceae (5%), Rubiaceae (4.5 %) and Moraceae (4 %). As a whole the tree density and basal area of 794 plants/ha and 71.05m²/ha, respectively are well within the reported range of various Indian tropical forests (Visalakshi, 1995; Sapkota et al., 2009).

3.6 Diversity measures

Species diversity, concentration of dominance and some mathematical indices of different vegetational layers of the reserve are given in Table -11. Measurement of biodiversity of specific area (local scale) on the basis of species richness does not provide a complete understanding about the individuals of the species in an ecosystem as it suffers from the

Plant group	Range of diversity indices			
	SD	CD	SR	SE
Tree	1.8 -3.11	0.07 - 0.316	3.36 - 6.59	0.611 - 0.951
Climber	0.63 - 1.86	0.155 - 0.36	0.55 - 2.22	0.8 - 0.931
Shrub	1.76 - 2.37	0.102 - 0.216	1.66 - 2.92	0.76 - 0.96
Herb	1.57 - 2.99	0.053 - 0.323	1.24 - 4.24	0.63 - 0.91
Sapling	2.1 - 3.03	0.061 - 0.194	2.98 - 6.15	0.816 - 0.949
Seedling	1.01 - 2.62	0.129 - 0.397	0.73 - 4.37	0.61 - 0.88

Table 11. Species diversity (SD), Concentration of dominance (CD), species richness (SR) and species evenness (SE) of different forest strata of Similipal biosphere reserve.

lack of evenness or equitability. It was observed that the richness index ranged from 3.36 to 6.59 (tree layer), 0.55 to 2.22 (climber layer), 1.24 to 4.24 (herb layer), 1.66 to 2.92 (shrub layer), 2.98 to 6.15 (sapling layer) and 0.73 to 4.37 (seedling layer). The equitability showed little variation across the vegetational layers which ranged from 0.61 to 0.95 (tree layer), 0.8 to 0.93 (climber layer), 0.63 to 0.91 (herb layer), 0.76 to 0.96 (shrub layer), 0.82 to 0.95 (sapling layer) and 0.61 to 0.88 (seedling layer). Shannon Wiener's index of diversity is one of the popular measures of species diversity. It ranged from 1.80 to 3.11, 0.63 to 1.86, 1.76 to 2.37, 1.57 to 2.99, 1.01 to 2.62 and 2.1 to 3.03 for tree, climber, shrub, herb, seedling and sapling layers, respectively, across all sites. Maximum range of species diversity of 1.8 to 3.11 was experienced by tree layer and the minimum range of 0.63 to 1.86 by climber layer indicating that tree layer of SBR was highly diverse while climber layer was the least (Table-11).

4. Discussion

4.1 Floristic composition

The species richness of a forest ecosystem depends on the number of species per unit area; the more species per unit area, the higher the species richness. A total of 266 species/ 3.6 ha or 74 species/ha in the SBR is more or less similar compared to the number of species reported by several workers in other tropical forest covers of India (Parthasarathy, 1999, 80 to 85 species/ ha in kalakad-MundanthuraiTiger reserve; Parthasarathy and Karthikeyan, 1997, 57 species/ ha in Mylodai- Courtallum reserve forest) and also 70 to 80 species/ha that have been observed in other studies in West African tropical high forests (Lawson, 1985; Vordzogbe et al., 2005). The species richness in neotropical forests showed a wide variation, ranging from 20 species/ha in Varzea forest of Rio Xingu, Brazil (Campbell et al., 1992) to 307 species/ha in the Amazonian Equator (Valenica et al., 1994). In the old world tropics species richness ranged from 26 species/ha in Kolli hills of India (Chittibabu and Parthasarathy, 2000) to 231 species/ha in Brunei Darussalam of South East Asia (Poulsen et al., 1996). In tropical rain forests the range of species count per hectare is about 20 to a maximum of 223. The number of species in SBR was 74 per hectare and this number is at the lower side of the range given in tropical rain forests and neotropical forests. In the study of species richness of the western ghat, south India Sunderpandian and Swamy (2000) stated that pronounced dry season and relatively low annual precipitation factors might be correlated with low species richness.

4.2 Diversity and related measurements

It has become common practice in quantitative descriptive studies to use IVI, which combines the relative frequency, density and dominance into a single measure to analyze a plant community. Though vegetation can be described in terms of a number of parameters including frequency, density and cover, the use of any one of these quantitative parameters could lead to over-simplification or under-estimation of the status of the species (Kigomo et. al., 1990, Oyun et. al., 2009). Except few tree species viz. *Shorea robusta*, *Terminalia alata* and *Anogeissus latifolia*, low ecological status of most of the tree species in the present investigation, as evidenced by the IVIs, may be attributed to lack of dominance by any one of these species, suggesting positive interactions among the tree species. In other words, resource spaces are shared to minimize negative species interactions and plants can obtain resources with relative ease (Tsingalia, 1990). The low IVIs may also imply that most of the tree species in this forest are rare (Pascal and Pellissier, 1996; Oyun et al., 2009), as confirmed by Raunkiaer's frequency distribution of the tree species (Table- 9). The rarity of species may be attributed to the occurrence of abundant sporadic species with low frequency in the stands (Oyun et al., 2009). The high percentage (>70 %) of rare species observed in various vegetational layers of the reserve confirms the generally acclaimed notion that most of the species in an ecological community are rare, rather than common (Magurran and Henderson, 2003). The range of evenness value and Simpson's diversity index of 0.61-0.96 and 0.053-0.397, respectively in vegetation layers of Similipal implies that most of the species are equitably distributed while very few species showed the degree of dominance (Pascal and Pellissier, 1996). Shannon Wiener species diversity value among vegetational layers of the reserve ranges from 0.63-3.11 indicating that SBR is highly diverse. The species diversity is generally higher for tropical forests, which is reported as 5.06 and 5.40 for young and old stand, respectively (Knight, 1975). For Indian forests the diversity index ranges between 0.83- 4.1 (Visalakshi, 1995). The diversity index of different vegetational layers of SBR is well within the reported range of the forests of Indian sub-continent (Table-11). Higher species diversity index in tropical forests as reported by Knight (1975) in comparison to the present investigation may be due to differences in the area sampled and lack of uniform plot dimensions. On the other hand, the value obtained for the concentration of dominance for vegetation layers of SBR (0.053-0.397) is greater than those recorded in Nelliampathy (0.085; Chandrashekara and Ramakrishnan, 1994) and tropical dry deciduous forests of Western India (0.08- 0.16; Nirmal Kumar et al., 2010). The high dominance value in the present study indicates single species dominance by *Shorea robusta* in tree, sapling and seedling layers of the reserve (Table-3, 7 and 8).

4.3 Distribution pattern of climbers

The distribution of climbers on the trees of the reserve was considerably low, being nearly equal to 14%. This may be due to high canopy coverage, thereby allowing low light to reach the forest floor and not triggering vigorous growth of climbers (Babweteera et al., 2001). The impact of climbers on the vitality of trees is negative (Toledo- Aceves and Swaine, 2008) causing loss of foliage and thereby reducing the surface area available for metabolic processes and reproductive potential as well as impeding or obstructing forest succession (Toledo-Aceves and Swaine, 2008). Notwithstanding their negative impacts, climbers form bridges between the forest canopies, thereby facilitating the movement of arboreal animals across the forest. They also protect weaker trees from strong winds (Schnitzer and Bongers, 2002).

4.4 Stand structure

Stand structure parameters allow predictions of forest biomass and can provide spatial information on potential determinants of plant species distributions (Couteron et al., 2005). In the present study stand structure relates to the basal area of trees, density of trees, and densities of herbs, shrubs, climbers, saplings and seedlings. The tree basal area of 71.05m²/ha is high and comparable to the reported range of various Indian tropical forests

(Visalakshi, 1995; Sapkota et al., 2009) and slightly higher than the value reported from Monteverde of Costa Rica (62 m²/ha, Nadkarni et al., 1995). High basal area is a characteristic feature of mature forest stand and serves as a reflection of high performance of the trees. It may also presuppose the development of an extensive root system used efficient nutrient absorption, growth suppressing of subordinate plants as the big trees intercept much of the solar radiation that might otherwise reach the forest floor. Dipterocarpaceae had the highest basal area in the present study, followed by Combretaceae, Myrtaceae, Rubiaceae and Moraceae. These families contain important timber species such as *Shorea robusta*, *Terminalia alata*, *Anogeissus latifolia*, *Syzygium cumini*, *Syzygium cerasoides*, *Terminalia bellirica*, *Terminalia chebula*, etc. The Barringtoniaceae, Chochlospermaceae, Clusiaceae, Malvaceae, Melastomataceae, Myrsinaceae, Ochnaceae, Rosaceae, Salicaceae, Lauraceae, Rhamnaceae, Sterculiaceae, Symplocaceae, Verbenaceae, Flacourtiaceae, and Rutaceae did not contribute much to the total basal area. In all the stands investigated, Chochlospermaceae, Sapotaceae, Salicaceae, etc. were by one individual each while Barringtoniaceae represented by two individuals and, Clusiaceae and Melastomataceae by three individuals each. This implies that very low contribution of these families to the total basal area may be due to their low numbers. Thus, these families may not be very important in terms of dominance. Species wise density of individuals having ≥30cm girth of the reserve ranged from less than one plant per hectare to 284 plants/ha and the total density of the reserve was 794 plants/ha. Maximum density (per hectare of individuals of ≥30cm) was recorded for *Shorea robusta* (284) followed by *Terminalia alata* (50), *Anogeissus latifolia* (45), *Protium serratum* (32) and *Dillenia pentagyna* (29). Density was observed less than or equal to one for many species like *Antidesma acidum*, *Artocarpus lacuccha*, *Butea monosperma,*, *Casearia elliptica*, *Chionanthus intermedicus*, *Cochlospermum religiosum*, *Euonymus glaber*, etc. Some other species showed intermediate range of density per hectare. The tree density of 794 individuals/ha recorded in the present investigation is lower as compared to densities reported from Saddle Peak of North Andaman Islands and Great Andaman Groups (946-1137 trees/ha, Padalia et al.,2004). However, the tree density of SBR is comparable with other tropical forests e.g. Kalkad Western Ghats (575-855 trees/ha, Parthasarathy, 1999), Brazil (420-777 trees/ha, Campbell et al., 1992), seasonally deciduous forest of Central Brazil (734 trees/ha, Felfili et al., 2007), Semideciduous forest of Piracicaba, Brazil (842 trees/ha, Viana and Tabanez, 1996) and Costa Rica (617 trees/ha, Heaney and Proctor, 1990). There appears to be little literature available to compare the herb, shrub, sapling and seedling densities with at the local level. The reported densities of these vegetation layers of the reserve in the present investigation is well comparable to Mishra et al. (2008). The fewer numbers of saplings recorded in relation to seedlings in this study implies that most of the saplings are transiting into young trees. It could also mean that most of the seedlings probably die due to intense competition (Weidelt, 1988) for available resources before they reach the sapling stage. Nevertheless, the totality of saplings and seedlings is colossal and reflects high regeneration potential of the forest (Mishra et al., 2005; Khumbongmayum et al., 2006).

4.5 Comparative analysis of tree species diversity in various tropical forests

The tree diversity observed in various tropical forests has also been compared with the findings of the present study in SBR (Table-12). The species diversity in SBR can be comparable with other tropical forests. Species richness and density of tree species of the present study (117 species and 794 plants per hectare) is well within the reported range of tropical forests in India and outside India. However, the basal area estimated for tree species in the present investigation is well within the reported range of Indian tropical forests but higher than that of tropical forests found outside India (Table-12). The high basal area of 71 m²/ha obtained in the present investigation was largely due to the contribution of the dominant tree species of the reserve, maximum by *Shorea robusta* which alone scored 39% (27.73 m²/ha) of basal area.

Forest and location	No. of species	No. of genera	No. of families	Density (Plants/ha)	Basal area (m2/ha)	Source
Indian tropical forest						
Moist deciduous forest, Similipal	117	87	42	793.67	71.04	Present study
Moist deciduous forest, Andaman	235	153	73	946	28.60	Padalia et al., 2004
Semi evergreen forest, Andaman	231	153	71	1027	33.76	Padalia et al., 2004
Evergreen forest, Andaman	264	176	81	1137	44.28	Padalia et al., 2004
Wet evergreen forest, South western ghat	122	89	41	575-855	61.7-94.64	Parthasarathy, 1999
Tropical forests outside India						
Neotropical cloud forest, Monteverde, Costarica	114	83	47	555	62.0	Nadkarni et al., 1995
Seasonally deciduous forest,Iaciara, Brazil	39	-	-	734	16.73	Felfili et al, 2007
Seasonally deciduous forest, Monte Alegre, Brazil	56	-	-	633	19.36	Nascimento et al., 2004
Semideciduous, Piracicaba, SP, Brazil	101	-	-	842	12.53	Viana and Tabanez, 1996
Evergreen rain forest, Ngovayang (Cameroon)	99-121	-	-	451-634	28.8-42.1	Christelle et al., 2011
Subtropical forest, Bagh district, Kashmir, Pakistan	72	-	31	344	69.31	Saheen et al., 2011
African wet tropical forest	344-494	-	-	371-486	27.8-35.8	Chuyong et al., 2011

Table 12. Tree species diversity in various tropical forests.

5. Conclusion

The overall analysis indicates that species rich communities of the moist deciduous tropical forests are not only being reduced in area but they are also becoming species poor and less diverse due to rapid deforestation and forest fragmentation. The community organization is also changing in response to increased anthropogenic disturbance. The study has shown that SBR is highly rich in plant diversity and is one of the treasure houses of good ecological wealth of Eastern ghat, India. The long history of timber exploitation prior to its conversion into a biosphere reserve has resulted in the alteration of structure of the forest whereby most of the tree species were affected by very few individuals. The ecological importance of most of the tree species was also low, which reflected rarity of most of the species. However, the abundance of small trees coupled with the colossal sum of saplings and seedlings reflects a high regeneration potential of the forest.

The forest management issues of SBR could be addressed by collection and analysis of long term ecological data which requires scientific baseline studies. We have covered extensively structural parameter analysis which is helpful to know the present state of ecological health of the ecosystem. But due to the various forms of anthropogenic pressure the habitat is destroyed with for logging, illegal hunting, and other challenges (mining in periphery, etc.). The conservation efforts have not so far yielded desired result. With continued biotic pressure and consequent change in structure and function of ecosystem, the management methodology also needs to be modified developing a Long Term Research Network. Similipal is a globally recognized ecosystem covered under UNESCO's Biosphere Reserve housing wide range of flora and fauna. We need to carry out Research and education activities to create an institutional platform to academicians, researchers and scientists. This ecosystem is under pressure. Continued destruction of old-growth and pristine forests of Similipal with high biodiversity will have a regional impact on social and ecological sustainability.

The over exploitation of natural resources in tropical world for meting the basic needs of food, fodder and shelter of local population has disturbed the landscapes causing rapid depletion of biodiversity. Our research results may be of some help to develop management schemes for conservation of biodiversity of SBR. Lack of data base and structural and functional characters of the ecosystem at regular intervals will not help to develop a long-term strategy for sustainable development. Thus continuous collection of data as per long-term action plan on successional status of species level up external and local pressures on the ecosystem, soil fertility management and linkage between social and ecological processes is needed. The community participation and use of traditional technologies as tools for natural resource management should be integrated to achieve sustainable resource management and ecological rehabilitation.

6. Acknowledgement

We are thankful to the Director, Dy. Director and other forest officials of Similipal biosphere reserve for their active cooperation in the field work. Financial assistance from CSIR and DST, New Delhi is greatly acknowledged.

7. References

Airi, S., Rawal, R. S., Dhar, U. and Purohit, A. N. (2000). Assessment of Availability and Habitat Preference of Jatamasi a Critically Endangered Medicinal Plant of West Himalayas. *Current Science*, Vol.79, pp.1467-1470, ISSN 0011-3891.

Alder, D. and Synnott, T. J. (1992). *Permanent Sample Plot Techniques for Mixed Tropical Forest.* Tropical Forestry Papers 25, ISBN 085074119, University of Oxford, UK.

Babweteera, F., Plumptre, A. and Obua, J. (2001). Effect of Gap Size and Age on Climber Abundance and Diversity in Budongo Forest Reserve, Uganda. *African Journal of Ecology*, Vol. 38, pp. 230-237, ISSN 0141-6707.

Campbell, D.G., Stone, J.L. and Rosas, A. Jr. (1992). A Comparison of the Phytosociology and Dynamics of Three Floodplain (Varzea) Forest of Known Ages, Rio Jurua, Western Brazilian Amazon. *Botanical Journal of the Linnean Society*, Vol. 108, pp. 231-237, ISSN 0024-4074.

Campbell, E. J. F. and Newbery, D. McC. (1993). Ecological Relationships Between Lianas and Trees in lowland Rain Forest in Sabh, East Malaysia. *Journal of Tropical Ecology*, Vol. 9, pp. 469-490, ISSN 0266-4674.

Champion, H.G. and Seth, S.K. (1968). *A Revised Survey of the Forest Types of India.* Manager of Publication, ISBN 81-8158-0613, New Delhi.

Chandrashekara, U. M. and Ramakrishnan, P. S. (1994). Vegetation and Gap Dynamics of a Tropical Wet Evergreen Forest in the Western Ghats of Kerala, India. *Journal of Tropical Ecology*, Vol. 10, pp. 337-354, ISSN 0266-4674.

Chittibabu, C.V. and Parthasarathy, N. (2000). Attenuated Tree Species Diversity in Human-Impacted Tropical Evergreen Forest Sites at Kolli Hills, Eastern Ghats, India. *Biodiversity and Conservation*, Vol. 9, pp. 1493-1519, ISSN 0960-3115.

Christelle, F. G., Doumunge, C., Mc Key, D., Tchouto, P. M. G., Sunderland, T. C. H., Balinga, M. P. B. and Snoke, B. (2011). Tree Diversity and Conservation value of Ngovayang's Lowland Forests, Cameroon. Biodiversity and Conservation, DOI 10.10007/s10531-011-0095-z, ISSN 0960-3115.

Chuyong , G. B., Kenfack, D., Harms, K. E., Thomas, D. W., Condit, R. and Comita, L. S. (2011). Habitat Specificity and Diversity of Tree Species in an African Wet Tropical Forest. *Plant Ecology*, Vol. 212, pp. 1363-1374, ISNN 1385-0237.

Couteron, P., Pelissier, R., Nicolini, E. A. and Paget, D. (2005). Predicting Tropical Forest Stand Parameters from Fourier Transform of Very High-Resolution Remotely Sensed Canopy Images. *Journal of Applied Ecology*, Vol. 42, pp. 1121-1128, ISSN 0021-8901.

Dixit, A.M. (1997). Ecological Evaluation of Dry Tropical Forest Vegetation: An Approach to Environmental Impact Assessment. *Tropical Ecology*, Vol. 38, pp. 87-100, ISSN 0564-3295.

FAO (2001). Food and Agriculture organization. *State of The Worlds Forests.* ISBN 92-5-104590-9, Rome.

Felfili, J., Nascimento, A. R. T., Fagg, C. W. and Meirelles, E. A. (2007). Floristic Composition and Community Structure of a Seasonally Deciduous Forest on Limestone Outcrops in Central Brazil. Revista Brasileria de Botanica, Vol. 30, No.4, pp. 611-621, ISSN 0100-8404.

Haines, H.H. (1925). *The Botany of Bihar and Orissa*. Vol. I-III. London, Botanical Survey of India, Calcutta (Repn.Edn. 1961).

Hare, M. A., Lantage, D. O., Murphy, P. G. and Checo, H. (1997). Structure and Tree Species Composition in a Subtropical Dry Forest in the Dominican Republic: Comparison with a Dry Forest in Puerto Rico. *Tropical Ecology*, Vol. 38, No. 1, pp. 1-17, ISSN 0564-3295.

Heaney, A. and Proctor, J. (1990). Preliminary Studies of Forest Structure and Floristics on Volcan Barva, Costa Rica. *Journal of Tropical Ecology*, Vol. 6, pp. 307-320, ISSN 0266-4674.

Heinrich, A. and Hurka, H. (2004). Species Richness and Composition During Sylvigenesis in a Tropical Dry Forest in North Western Costa Rica. *Tropical Ecology*, Vol.45, No. 1, pp. 43-57, ISSN 0564-3295.

Hewit, N. and Kellman, M. (2002). True Seed Dispersal Among Forest Fragments: Dispersal Ability and Biogeographical Controls. *Journal of Biogeography*, Vol. 29, No. 3, pp. 351-363, ISSN 0305-0270.

Kershaw, K.R. (1973). *Quantitative and Dynamic Plant Ecology*. Edward Arnold Ltd., ISBN 0713120991, London.

Khumbongmayum, A.D., Khan, M. L. and Tripathi, R. S. (2006). Biodiversity Conservation in Sacred Groves of Manipur, Northeast India: Population Structure and Regeneration Status of Woody Species. *Biodiversity and Conservation*, Vol.15, pp. 2439-2456, ISSN 0960-3115.

Kigmo, B. N., Savill, P. S. and Woodell, S. R. (1990). Forest Composition and Its Regeneration Dynamics: A Case Study of Semi-Deciduous Tropical Forest in Kenya. *African Journal of Ecology*, Vol. 28, No. 3, pp. 174-187, ISSN 0141-6707.

Knight, D.H. (1975). A Phytosociological Analysis of Species Rich Tropical Forest on Barro Colorado Island, Panama. *Ecological Monographs*, Vol. 45, pp. 259-289, ISSN 0012-9615.

Krishnankutty, N., Chandrasekaran, S. and Jeyakumar, G. (2006). Evaluation of Disturbance in a Tropical Dry Deciduous Forest of Alagar Hill (Eastern Ghats), South India. *Tropical Ecology*, Vol. 47, No. 1, pp. 47-55, ISSN 0564-3295.

Lawson, G. W. (1985). *Plant Life in West Africa*. Ghana Universities Press, ISBN 9964-3-0118-9, Accra.

Lewis, O. T. (2009). Biodiversity Change and Ecosystem Function in Tropical Forests. *Basic and Applied Ecology*, Vol.10, No. 2, pp. 97-102, ISSN 1439-1791.

Ludwig, J. A. and Reynolds, J. F. (1988). *Statistical Ecology: A Primer on Methods and Computing*. Wiley-Interscience, ISBN 10-0471832359, New York.

Magurran, A. E. and Henderson, P. A. (2003). Explaining the Excess of Rare Species in Natural Species Abundance Distributions. *Nature*, Vol. 422, pp. 714-716, ISSN 0028-0836.

Mayers, N. (1992). *The Primary Source: Tropical Forests and Our Future*. Norton and Co., ISBN 978-0-393-30828-0, New York.

Mishra, B. P., Tripathi, O. P. and Laloo, R. C. (2005). Community Characteristics of a Climax Subtropical Humid Forest of Meghalaya and Population Structure of

Ten Important Tree Species. *Tropical Ecology*, Vol. 46, pp. 241-251, ISSN 0564-3295.

Mishra, R. K., Upadhyay, V. P. and Mohanty, R. C. (2008): Vegetation Ecology of the Similipal Biosphere Reserve, Orissa, India. *Journal of Applied Ecology and Environmental Research*, Vol.6, No.2, pp. 89-99, ISSN 1589-1623.

Mishra, R. K., Upadhyay, V. P., Mohapatra, P. K., Bal, S. and Mohanty, R. C. (2006). Phenology of Species of Moist Deciduous Forest Sites of Similipal Biosphere Reserve. *Lyonia*, Vol.11, No.1, pp.5-17, ISSN 0888-9619.

Mueller-Dombois, D. and Ellenberg, H. (1974). *Aims and Methods of Vegetation Ecology*. John Wiley and Sons, ISBN 0471622915, New York.

Murthy, M. S. R., Sudhakar, S., Jha, C. S., Reddy, S., Pujar, G. S. and Roy, P. S. (2007). Vegetation, Land Cover and Phytodiversity Characterisation and Landscape Level Using satellite Remote Sensing and Geographic Information System in Eastern Ghats, India. *EPTRI-ENVIS Newsletter*, Vol. 13, No. 1, pp. 1-12,ISSN 0974-2336.

Nadkarni, N. M., Matelson, T.S. and . Haber, W.A. (1995). Structural Characteristics and Floristic Composition of Neotropical Cloud Forest, Monte Verde, Costa Rica. *Journal of Tropical Ecology*, Vol.11, pp. 481-495, ISSN 0266-4674.

Nascimento, A. R. T., Felfili, J. M. and Meirelles, E. M. L. (2004). Floristica e Estrutura De Um Remanscent De Floresta Estucional Decidual De Encosta no Municipio De Monte Alegre, GO, Brazil. *Acta Botanica Brasilica*, Vol. 18, pp. 659-699, ISSN 0102-3306.

Newbery, D. McC. (1991). Floristic Variation Within Kerangas (Heath) Forest: revaluation of Data from Sarawak and Brunei. Vegetatio, Vol. 96, pp. 43-86, ISSN 0042-3106.

Newbery, D. McC., Kennedy, D. N., Petol, G. H., Madani, L. and Ridsdale, C. E. (1999). Primary Forest Dynamics in Lowland Dipterocarp Forest at Danum Valley, Sabh, Malaysia and the Role of Understorey. *Philosophical Transactions of Royal Society of London*, Series B, 354, pp.1763-1782, ISSN 0080-4622.

Nirmal Kumar, J. I., Kumar, R. N., Bhoi, R. K. and Sajish, P. R. (2010). Tree Species Diversity and Soil Nutrient Status in Three Sites of Tropical Dry Deciduous Forest of Western India. Tropical Ecology, Vol. 51, No. 2, pp. 273-279, ISSN 0564-3295.

Odum, E.P. (1971). *Fundamentals of Ecology*. W.B. Saunders Co., ISBN 0-943451-25-6, Philadelphia.

Oyun, M. B., Bada, S. O. and Anjah, G. M. (2009). Comparative Analysis of the Floral Composition at the Edge and Interior of Agulii Forest Reserve, Cameroon. *Journal of Biological Science*, Vol.9, No.5, pp.431-437, ISSN 1727-3048.

Padalia, H., Chauhan, N., Porwal, M.C. and Roy, P. S. (2004). Phytosociological Observations on Tree Species Diversity of Andaman Islands, India. *Current Science*, Vol. 87, pp. 799-806, ISSN 0011-3891.

Parker, J. S. (1984). UNESCO Documents and Publications in the Field of Information: A Summary Guide. IFLA Journal, Vol. 10, No.3, pp. 251-272, ISSN 0340-0352.

Parthasarathy, N. (1999). Tree Diversity and Distribution in Undisturbed and Human-Impacted Sites of Tropical Wet Evergreen Forest in Southern Western Ghats, India. *Biodiversity and Conservation*, Vol. 8, pp. 1365-1381, ISSN 0960-3115.

Parthasarathy, N. and Karthikeyan, R.(1997). Biodiversity and Population Density of Woody Species in a Tropical Evergreen Forest in Courtallum Reserve Forest, Western Ghats, India. *Tropical Ecology*, Vol. 38, pp. 297-306, ISSN 0564-3295.

Pascal L. P., Pellissier, R. (1996). Structure and Floristic Composition of a Tropical Evergreen Forest in South-West India. *Journal of Tropical Ecology*, Vol.12, pp. 191- 210, ISSN 0266-4674.

Pielou, E.C. (1975). *Ecological Diversity*. John Wiley and Sons, ISBN 0471689254, New York.

Poulsen, A.D., Nielsen I.V. and Balslev, H. (1996). A Quantitative Inventory of Trees in One Hectare of Mixed Dipterocarp Forest in Temburong, Brunei Durussalam, In *Tropical Rain Forest Research-Current Issues*, D.S. Edwards, W.E. Booth and S.C. Choy (Eds.) Kluwer Academic Publishers, ISBN 0-7923-4038-8, Dordrecht, The Netherlands.

Puri, G. S. (1995). Biodiversity and Development of Natural Resources for the 21st Century. *Tropical Ecology*, Vol. 56, No. 2, pp. 253-255, ISSN 0564-3295.

Raunkiaer, C. (1934). *The Life Form of Plants and Statistical Plant Geography*. Claredon Press ISBN 9978-40-943-2, Oxford.

Reddy, C. S., Pattanaik, C., Mohapatra, A. and Biswal, A. K. (2007). Phytosociological Observations on Tree Diversity of Tropical Forest of Similipal Biosphere Reserve, Orissa, India. *Taiwania*, Vol. 52, No.4, pp. 352-359, ISSN 0065-1125.

Rout, S. D., Panda, T. and Mishra, N. (2009). Ethnomedicinal Studies on Some Pteridophytes of Similipal Biosphere Reserve, Orissa, India. *International Journal of Medicine and Medical Sciences*, Vol.1, No.5, pp. 192-197, ISSN 2006-9723.

Rout, S. D., Panda, S. K., Mishra, N. and Panda, T. (2010). Role of Tribals in Collection of Commercial Non-Timber Forest Products in Mayurbhanj District, Orissa. *Journal of Studies of Tribes and Tribals*, Vol.8, No. 1, pp. 21-25, ISSN 0972-639X.

Roy, P. S., Dutt, C. B. S. and Joshi, P. K. (2002). Tropical Forest Resource Assessment and Monitoring. *Tropical Ecology*, Vol. 43, No. 1, pp. 21-37, ISSN 0564-3295.

Sapkota, I.P., Tigabu, M. and Oden, P. C. (2009). Species Diversity and Regeneration of Old-growth Seasonally Dry *Shorea robusta* Forests Following Gap Formation. *Journal of Forestry Research*, Vol. 20, pp. 7-14, ISSN 1007-662X.

Saxena, H.O. and Brahmam, M. (1989): *The Flora of Similipahar (Similipal)*. Orissa. Regional Research Laboratory, Bhubaneswar.

Saxena, H.O. and Brahmam, M. (1996). *The Flora of Orissa*. Vol. I-IV, Regional Research Laboratory (CSIR), Bhubaneswar and Orissa Forest Development Corporation Ltd., Bhubaneswar.

Schnitzer, S. A. and Bongers, F. (2002). The Ecology of Lianas and Their Role in Forests. *Trends in Ecology and Evolution*, Vol.17, No.5, pp. 223-230, ISSN 0169-5347.

Shaheen, H., Qureshi, R. A. and Shinwari, Z. K. (2011). Structural Diversity, Vegetation Dynamics and Antropogenic Impact on Lesser Himalayan Subtropical Forests of Bagh District, Kashmir. *Pakistan Journal of Botany*, Vol. 43, No. 4, pp. 1861-1866, ISSN 0556-3321.

Shannon, C.E. and Wiener, W. (1963). *The Mathematical Theory of Communication*. University Press, ISBN 0252725484, Illinois, USA.

Simpson, E.H. (1949).   Measurement of Diversity. *Nature*, Vol. 163, pp. 688, ISSN 0028-0836.

Singh, J.S. (2002). The Biodiversity Crisis: A Multifaceted Review. *Current Science*, Vol. 82, pp. 638-647, ISSN 0011-3891.

Srivastava, S. S. and Singh, L. A. K. (1997). Monitoring of Precipitation and Temperature of Similipal Tiger Reserve, In: *Similipal: A National Habitat of Unique Biodiversity*, P. C. Tripathy and S. N. Patro (Eds.), 30-40, Orissa Environmental Society, ISBN 81-900256-5-1, Bhubaneswar.

Sunderpandian, S.M. and Swamy, P.S. (2000). Forest Ecosystem Structure and Composition Along an Altitudinal Gradient in the Western Ghats, South India. *Journal of Tropical Forest Science*, Vol.12, No. 1, pp. 104-123, ISSN 0128-1283.

Tilman, D. (2000). Causes, Consequences and Ethics of Biodiversity. *Nature*, Vol. 405, pp. 208-211, ISSN 0028-0836.

Toledo-Aceves, T. and Swaine, M. D. (2008). Effect of Lianas on Tree Regeneration in Gaps and Forest Understorey in a Tropical Forest in Ghana. *Journal of Vegetation Science*, Vol. 19, No. 5, pp. 717-728, ISSN 1100-9233.

Townsend, A. R., Asner, G. P. and Cleveland, C. C. (2008). The Biogeochemical Heterogeneity of Tropical Forests. *Trends in Ecology and Evolution*, Vol. 23, No. 8, pp. 424-431, ISSN 0169-5347.

Tripathi, K. P. and Singh, B. (2009).Species Diversity and Vegetation Structure Across Various Strata in Natural and Plantation Forest in Katerniaghat Wildlife Sanctuary, North India. *Tropical Ecology*, Vol. 50, No. 1, pp. 191-200, ISSN 0564-3295.

Tsingalia, M. T. (1990). Habitat Disturbance, Severity and Patterns of Abundance in Kakamega Forest, Western Kenya. *African Journal of Ecology*, Vol. 28, pp. 213- 226, ISSN 0141-6707.

Valencia, R., Balslev, H. and Mino, P. (1994). High Tree Alpha Diversity in Amazonian Ecuador. *Biodiversity and Conservation*, Vol. 3, pp. 21-28, ISSN 0960-3115.

Vasanthraj, B. K., Shivaprasad, P. V. and Chandrashekar, K. R. (2005). Studies on the Structure of Jadkal Forest, Udupi District, India. *Journal of Tropical Forest Science*, Vol. 17, pp. 13-32, ISSN 0128-1283.

Viana, V.M. and Tabanez, A.A.J. (1996). Biology and Conservation of Forest Fragments in the Brazilian Atlantic Moist Forest. In: *Forest Patches in Tropical Landscapes*, J. Schelas and R. Greenberg (Eds.) 151-167. Island Press, ISBN 1-55963-425-1, Washington.

Visalakshi, N. (1995). Vegetation Analysis of Two Tropical Dry Evergreen Forests in Southern India. *Tropical Ecology*, Vol. 36, pp. 117-127, ISSN 0564-3295.

Vordzogbe, V. V., Attuquayefio, D. K. and Gbogbo, F. (2005). The Flora and Mammals of Moist Semi-Deciduous Forest Zone in the Sefwi Wiawso District of the Western Region, Ghana. *African Journal of Ecology*, Vol. 8, pp. 49-64, ISSN 0266-4674.

WCMC (1992). World Conservation Monitoring Centre. *Global Biodiversity: Status of Earth's Living Resources*. Chapman and Hall, ISBN 0412-47240-6, London, UK.

Weidelt, H. J. (1988). On the Diversity of Tree Species in Tropical Rain Forest Eosystems. *Plant Research and Development*, Vol. 28, pp. 110-125, ISSN 0018-8646.

Whitford, P.B. (1949). Distribution of Woody Plants in Relation to Succession and Clonal Growth. *Ecology*, Vol. 30, pp. 199-208, ISSN 0012-9658.

Permissions

The contributors of this book come from diverse backgrounds, making this book a truly international effort. This book will bring forth new frontiers with its revolutionizing research information and detailed analysis of the nascent developments around the world.

We would like to thank Dr. Juan A. Blanco and Dr. Yueh-Hsin Lo, for lending their expertise to make the book truly unique. They have played a crucial role in the development of this book. Without their invaluable contribution this book wouldn't have been possible. They have made vital efforts to compile up to date information on the varied aspects of this subject to make this book a valuable addition to the collection of many professionals and students.

This book was conceptualized with the vision of imparting up-to-date information and advanced data in this field. To ensure the same, a matchless editorial board was set up. Every individual on the board went through rigorous rounds of assessment to prove their worth. After which they invested a large part of their time researching and compiling the most relevant data for our readers. Conferences and sessions were held from time to time between the editorial board and the contributing authors to present the data in the most comprehensible form. The editorial team has worked tirelessly to provide valuable and valid information to help people across the globe.

Every chapter published in this book has been scrutinized by our experts. Their significance has been extensively debated. The topics covered herein carry significant findings which will fuel the growth of the discipline. They may even be implemented as practical applications or may be referred to as a beginning point for another development. Chapters in this book were first published by InTech; hereby published with permission under the Creative Commons Attribution License or equivalent.

The editorial board has been involved in producing this book since its inception. They have spent rigorous hours researching and exploring the diverse topics which have resulted in the successful publishing of this book. They have passed on their knowledge of decades through this book. To expedite this challenging task, the publisher supported the team at every step. A small team of assistant editors was also appointed to further simplify the editing procedure and attain best results for the readers.

Our editorial team has been hand-picked from every corner of the world. Their multi-ethnicity adds dynamic inputs to the discussions which result in innovative outcomes. These outcomes are then further discussed with the researchers and contributors who give their valuable feedback and opinion regarding the same. The feedback is then collaborated with the researches and they are edited in a comprehensive manner to aid the understanding of the subject.

Apart from the editorial board, the designing team has also invested a significant amount of their time in understanding the subject and creating the most relevant covers. They scrutinized every image to scout for the most suitable representation of the subject and create an appropriate cover for the book.

The publishing team has been involved in this book since its early stages. They were actively engaged in every process, be it collecting the data, connecting with the contributors or procuring relevant information. The team has been an ardent support to the editorial, designing and production team. Their endless efforts to recruit the best for this project, has resulted in the accomplishment of this book. They are a veteran in the field of academics and their pool of knowledge is as vast as their experience in printing. Their expertise and guidance has proved useful at every step. Their uncompromising quality standards have made this book an exceptional effort. Their encouragement from time to time has been an inspiration for everyone.

The publisher and the editorial board hope that this book will prove to be a valuable piece of knowledge for researchers, students, practitioners and scholars across the globe.

List of Contributors

Katarína Merganičová and Ján Merganič
Czech University of Life Sciences in Prague, Faculty of Forestry, Wildlife and Wood Sciences, Praha, Czech Republic
Forest Research, Inventory and Monitoring (FORIM), Železná Breznica, Slovakia

Miroslav Svoboda and Radek Bače
Czech University of Life Sciences in Prague, Faculty of Forestry, Wildlife and Wood Sciences, Praha, Czech Republic

Vladimír Šebeň
National Forest Centre, Forest Research Institute, Zvolen, Slovakia

Ján Merganič and Katarína Merganičová
Czech University of Life Sciences in Prague, Faculty of Forestry, Wildlife and Wood Sciences, Department of Forest Management, Praha, Czech Republic
Forest Research, Inventory and Monitoring (FORIM), Železná Breznica, Slovakia

Róbert Marušák and Vendula Audolenská
Czech University of Life Sciences in Prague, Faculty of Forestry, Wildlife and Wood Sciences, Department of Forest Management, Praha, Czech republic

Camila Maistro Patreze, Milene Moreira and Siu Mui Tsai
UNIRIO- Universidade Federal do Estado do Rio de Janeiro, APTA – Polo Centro Sul, Laboratório de Biologia Celular e Molecular, CENA – Universidade de São Paulo, Brazil

Pio Federico Roversi and Roberto Nannelli
CRA – Research Centre for Agrobiology and Pedology, Florence, Italy

Chimène Assi-Kaudjhis
Université de Versailles Saint Quentin-en-Yvelines, France

Rodolfo Martinez Morales
International Crops Research Institute for the Semi-Arid Tropics, Niamey, Niger

R.K. Mishra
Department of Wildlife and Biodiversity Conservation, North Orissa University, Takatpur, Baripada, India

V.P. Upadhyay
Ministry of Environment and Forests, Eastern Regional Office, Chandrasekharpur, Bhubaneswar, india

P.K. Nayak, S. Pattanaik and R.C. Mohanty
Department of Botany, Utkal University, Vani Vihar, Bhubaneswar, India